Report of Investigations 9684

Practical Demonstrations of Ergonomic Principles

Susan M. Moore, Ph.D., Janet Torma-Krajewski, Ph.D., C.I.H., C.P.E.,
Lisa J. Steiner, M.S., C.P.E.

DEPARTMENT OF HEALTH AND HUMAN SERVICES
Centers for Disease Control and Prevention
National Institute for Occupational Safety and Health
Pittsburgh Research Laboratory
Pittsburgh, PA

July 2011

This document is in the public domain and may be freely copied or reprinted.

Disclaimer

Mention of any company or product does not constitute endorsement by the National Institute for Occupational Safety and Health (NIOSH). In addition, citations to Web sites external to NIOSH do not constitute NIOSH endorsement of the sponsoring organizations or their programs or products. Furthermore, NIOSH is not responsible for the content of these Web sites.

The findings and conclusions in this report are those of the author(s) and do not necessarily represent the views of the National Institute for Occupational Safety and Health

Ordering information

To receive documents or other information about occupational safety and health topics, contact NIOSH at

Telephone: **1–800–CDC–INFO** (1–800–232–4636)
TTY: 1–888–232–6348
E-mail: cdcinfo@cdc.gov

or visit the NIOSH Web site at **www.cdc.gov/niosh**

For a monthly update on news at NIOSH, subscribe to NIOSH eNews by visiting **www.cdc.gov/niosh/eNews**.

DHHS (NIOSH) Publication No. 2011-191

July 2011

SAFER • HEALTHIER • PEOPLE™

CONTENTS

ACKNOWLEDGEMENTS v
GLOSSARY vi
SECTION 1: Introduction 1
    Target Users and Audiences 1
    Format of Demonstration Descriptions 2
    Suggested Supplies 2
    Graphics 3
SECTION 2: Neutral Postures 4
    Principles 19
SECTION 4: Hand-Tool Selection and Use 28
    Principles 28
SECTION 5: Fatigue Failure and Back Pain 37
    Principles 37
SECTION 6: Moment Arms and Lifting 41
    Principles 41
APPENDIX A: Suggested Supplies 47
APPENDIX B: Useful Images for Handouts 52
REFERENCES 56

FIGURES

Figure 1. Neutral and awkward wrist postures. 5
Figure 2. Neutral and awkward elbow postures. 6
Figure 3. Neutral and awkward shoulder postures. 6
Figure 4. Neutral and awkward back postures. 7
Figure 5. Example of a portable EMG device (showing electrodes on skin) that indicates muscle activity by emitting audible signals. 8
Figure 6. Wrist postures and electrode placement for portable EMG device. 9
Figure 7. Negative, flat, and positive tilt positions for a keyboard. 10
Figure 8. Electrode placement on the upper arm. 11
Figure 9. Electrode placement for the shoulder. 12
Figure 10. Neutral, abducted, and flexed (reaching) shoulder postures. 13
Figure 11. Electrode placement for the back (line indicates location of spine). It is important that the electrodes are placed on the muscles as shown in the photograph. If the electrodes are placed too high on the back, the demonstration will not work properly. 14
Figure 12. Neutral, moderately flexed, and highly flexed postures of the back. 15
Figure 13. Hand dynamometer showing how wrist angle affects force production for neutral, ulnar deviation, and radial deviation wrist postures. 18
Figure 14. Pinch (lateral) grip and power grip. 20
Figure 15. Example of the maximum forces generated for a pinch grip (lateral) and a power grip 22
Figure 16. Example of a pinch grip (lateral) and the resulting maximum force. 24
Figure 17. The power grip is shown for five different grip widths. The narrowest grip is Grip 1; the width increases for each subsequent grip, with Grip 5 being the widest grip. ...26
Figure 18. Maximum–force output for each grip width. Note that, for this participant, Grip 2 had the highest force production. 27
Figure 19. Evaluating the effect of tool-handle diameter. 30
Figure 20. Examples of situations in which a pistol grip and inline grip would be useful as a means for keeping the wrist in a neutral posture. 32
Figure 21. Two types of pliers, one with a spring that reduces forceful exertion when opening the jaw, and one without a spring. 34
Figure 22. Example of one-handed and two-handed drilling. 36
Figure 23. Image of vertebrae, disc, and endplates. 37
Figure 24. A pen cap that is bent multiple times visually shows fatigue; a paper clip shows the result of failure. The graph (generalized for bone) illustrates how the same load, lifted many times, may ultimately, over time, lead to failure. 40
Figure 25. These schematics illustrate how increasing the distance between the worker and the object being lifted increases the overall moment (i.e., torque) for which the back muscles must compensate by expending more force. 42
Figure 26. A moment-arm simulator showing that more force/weight (W; arrow indicates direction of force) is needed to balance the "see saw" if the moment arm (L) is shorter on one side of the fulcrum as compared to the other side. 44
Figure 27. Moment-arm simulator with dial scale showing that, as the moment arm is increased, the resulting force acting on the scale increases. 46
Figure A-1. Dimensions for the moment-arm simulator. 49
Figure A-2. Examples of the suggested supplies for the demonstrations. 51

ACKNOWLEDGEMENTS

The technical contributions of Dr. Sean Gallagher pertaining to methods and materials used for the demonstrations and his assistance writing the background for the lower back demonstrations (Section 5) are acknowledged. Additionally, the assistance of Jonisha Pollard and Mary Ellen Nelson in obtaining photographs of the demonstrations and the artistic work of Alexis Wickwire are also acknowledged. The authors also appreciate Alan Mayton's and Patrick McElhinney's willingness to participate in the demonstrations video. Videotaping and editing of the videos were professionally completed by Charles Urban.

GLOSSARY

Awkward posture. Deviation from the natural or "neutral" position of a body part. A neutral position places minimal stress on the body part. Awkward postures typically include reaching overhead or behind the head; twisting at the waist; bending the torso forward, backward, or to the side; squatting; kneeling; and bending the wrist.

Cumulative injury (overuse injury). Cumulative injuries develop from repeated loading of body tissues over time. Such injuries include overuse sprains/strains, herniated discs, tendonitis, and carpal tunnel syndrome.

Disorder. A medical condition that occurs when a body part fails to function properly.

Ergonomics. The science of fitting workplace conditions and job demands to the capabilities of workers, and designing and arranging items in the workplace for efficiency and safety.

Fatigue failure. The weakening or breakdown of material subjected to stress, especially a repeated series of stresses.

Force. The amount of physical effort a person uses to perform a task.

Inline grip. A hand tool with a straight handle that is parallel with the direction of the applied energy.

Moment (torque). The tendency to produce motion about an axis.

Moment arm. The perpendicular distance between an applied force and the axis of rotation. For muscles, this is the perpendicular distance between the line of action of the muscle and the center of rotation at the joint.

Musculoskeletal disorders (MSDs). Illnesses and injuries that affect one or more parts of the soft tissue and bones in the body. The parts of the musculoskeletal system are bones, muscles, tendons, ligaments, cartilage, and their associated nerves and blood vessels.

Neutral body posture. The resting position of body parts.

Pinch grip. A grasp in which one presses the thumb against the fingers of the hand and does not involve the palm.

Pistol grip. A tool handle that resembles the handle of a pistol and is typically used when the tool axis must be elevated and horizontal or below waist height and vertical.

Power grip. A grasp where the hand wraps completely around a handle, with the handle running parallel to the knuckles and protruding on either side.

Repetitive. Performing the same motions repeatedly over time. The severity of risk depends on the frequency of repetition, speed of the movement, number of muscle groups involved, and required force.

Risk factor. An action and/or condition that may cause an injury or illness, or make an existing injury or illness worse. Examples related to ergonomics include forceful exertion, awkward posture, and repetitive motion.

Stress. Demand (or "burden") on the human body caused by something outside of the body, such as a work task, the physical environment, work-rest schedules, and social relationships.

Traumatic injury. Injuries that are acute, that may result from instantaneous events such as being struck by objects and that often require immediate medical attention. These types of injuries are often sustained through accidents.

SECTION 1: INTRODUCTION

Musculoskeletal disorders (MSDs) often involve the back, wrist, elbow, and/or shoulder, and occur when workers are exposed over time to MSD risk factors, such as awkward postures, forceful exertions, or repetitive motions. These exposures sometimes occur due to poorly designed workstations, tasks, and/or hand tools [Chaffin et al. 2006; Sanders and McCormick 1993; Silverstein et al. 1996, 1997]. Workers must understand the nature of MSD risk factors and how to avoid exposure to them. In a classroom setting, trainers may discuss ergonomic principles and show examples of MSD risk factors with photographs or videos. However, supplementing training with practical, hands-on demonstrations may further reinforce these ergonomic principles and help workers understand the importance of avoiding exposure to MSD risk factors. Moreover, demonstrations that allow for worker participation result in a greater understanding of the impact exposures to particular MSD risk factors have on workers' bodies. This document consists of a series of demonstrations designed to complement training on ergonomic principles. A description of the materials needed and step-by-step methodology are included in this document. Each demonstration highlights worker participation and uses relatively inexpensive materials.

The demonstrations are organized by type of ergonomic principle. Five general topics are addressed:

- Neutral compared with non-neutral postures
- Grip types
- Hand-tool selection and use
- Fatigue failure and back pain
- Moment arms and lifting

The demonstrations show the effects of posture, work methods, workstation design, tools, tasks, and location of materials on worker exposure to MSD risk factors. Many of the demonstrations are appropriate supplements to the NIOSH-developed training "Ergonomics and Risk Factor Awareness Training for Miners," which is provided to mining employees and downloadable from the NIOSH mining website:
http://www.cdc.gov/niosh/mining/pubs/pubreference/outputid2748.htm [NIOSH 2008].

Target Users and Audiences

This document was developed for individuals who intend to provide training on ergonomic principles that focus on MSD risk-factor exposures. It was designed for trainers of all experience levels including the beginning trainer. The demonstrations are designed to be performed by both the trainer and the worker. Each demonstration reinforces specific ergonomic principles and teaches the worker how and why to avoid MSD risk factors. Additionally, individuals involved in the purchase and selection of new and/or replacement tools may benefit from many of the demonstrations because they highlight the importance of considering ergonomic principles before purchasing tools.

Format of Demonstration Descriptions

Each section of this document begins with a discussion of an ergonomic principle and its role in avoiding MSD risk factors, followed by a series of demonstrations that may be used to show how the principle can be incorporated into the work environment. Each demonstration starts with clear objective statements and concludes with take-home messages that participants should incorporate into their everyday thinking. These demonstrations encourage audience participation because discussing how the principle plays a role in a worker's specific workplace is important for promoting understanding. Each demonstration includes the following information:

- Objectives of the demonstration
- List of suggested supplies needed to conduct the demonstration
- Step-by-step demonstration methodology
- Take-home messages that should be emphasized during the demonstration

To assist the trainer in knowing how to use some of the suggested supplies (e.g., portable EMG device, hand dynamometer), a series of brief video clips on DVD are included with this document. The DVD also contains video clips that show how to perform the demonstration and the results you should receive when using the portable electromyography (EMG) device.

Suggested Supplies

A complete list of required supplies for performing the demonstrations is provided in Appendix A. As previously mentioned, each demonstration description includes a list of supplies specific to that demonstration. Most of the supplies are available at hardware stores for a reasonable cost.

A portable EMG device is recommended for use with several of the demonstrations (Please see Appendix A for more information regarding the purchase of this device including cost and potential manufacturers). An EMG device is a rudimentary instrument that can be used to make relative comparisons of muscle activity by measuring the electrical activity of a muscle. The muscle emits an electrical signal when it undergoes one of two types of contractions—concentric (i.e., when the muscle shortens as it contracts) or eccentric (i.e., when the muscle lengthens as it contracts). When a contraction of the muscle is detected, the device emits an audible signal of beeps; the frequency of these beeps increases as the measured activity of the muscles beneath the electrodes increases. For example, if you place the electrodes on the inside of the forearm and ask the participant to contract the forearm muscles to 50% of their maximal effort, you will hear the device beep at a specific frequency. If you then ask the participant to contract his or her forearms to a maximal level, the frequency of the beeps will increase.

One problem with measuring muscle activity using electrodes placed on the skin is that the electrodes may measure what is referred to as "crosstalk." Crosstalk is produced when an electrode measures a signal over a nonactive or nearby muscle. For example, if you apply the electrodes to the inside of the forearm of the participant and have the participant flex his or her wrist (i.e., move the palm towards the inside of the forearm), the forearm muscles being measured are contracting and the EMG device will emit the audible signal. However, if you have the participant extend his or her wrist (i.e., move the palm as far away from the inside of the forearm as they can), the EMG device will still emit an audible signal even though the muscles the electrodes reside above are not contracting. This is an example of crosstalk where the electrodes are detecting activity from the muscles on the other side of the forearm. The electrodes could also be detecting a small eccentric contraction if the participant is using the inner forearm muscles to control the rate at which the wrist is extended. If the trainer is not aware of these issues, the trainer and the participants may be confused by the seemingly mistaken readings. The trainer should practice the demonstrations provided in this document prior to attempting them in front of an audience to minimize any occurrences where this "crosstalk" could cause confusion for the audience members.

Another limitation in using electrodes is that the amount of electrical activity produced by the different muscles of the body varies. Therefore, the EMG device provides the user with the ability to select from several different scales. You may need to adjust the scale in order for the device to be sensitive enough to detect changes in muscle activity for your muscles of interest. You must use the same scale the entire time you are measuring the electrical activity of a specific muscle group; otherwise, you will not be able to make a direct comparison to the muscle activity before and after an event.

Graphics

Appendix B includes several images that may be useful to show in a PowerPoint presentation when conducting the demonstrations. Electronic files for these graphics are found on the DVD provided with this document.

SECTION 2: NEUTRAL POSTURES

> ### Principles
>
> - **Use neutral postures:**
>   - **Maximum muscle force producible in neutral postures is greater than maximum muscle force producible in awkward postures.**
>   - **Fatigue occurs sooner when working in awkward postures.**
>   - **Working in extreme awkward postures (near extreme ranges of motion) causes stress on muscles and joints.**

A neutral posture is achieved when the muscles are at their resting length and the joint is naturally aligned. For most joints, the neutral posture is associated with the midrange of motion for that joint. When a joint is not in its neutral posture, its muscles and tendons are either contracted or elongated. Joints in neutral postures have maximum control and force production [Basmajian and De Luca 1985; Chaffin et al. 2006]. Neutral postures also minimize the stress applied to muscles, tendons, nerves, and bones. A posture is considered "awkward" when it moves away from the neutral posture toward the extremes in range of motion.

For the most part, a worker is capable of producing his or her highest amount of force when a joint is in its neutral posture. As the joint moves away from the neutral posture, the amount of force the muscles can produce decreases because some of the muscle fibers are either contracted or elongated [Clarke 1966; Kumar 2004]. Also, when you bend your wrist, the tendons of the muscles partially wrap around the carpal bones in the wrist. Because the bones do not act as a perfect pulley, a loss in the force that can be produced will occur. Furthermore, losses in force are also experienced due to friction [Ozkaya and Nordin 1999]. Thus, in order for a worker to produce the same force in an awkward posture as they do in the neutral posture, the worker's muscles must work harder and expend more energy. Working in an awkward posture, therefore, is a MSD risk factor that should be avoided. This is an extremely important principle because working closer to one's maximum capability, especially without rest, may result in an earlier onset of fatigue and, over time, may also increase the risk of MSDs [Chaffin et al. 2006]. Ideally, tasks and workspaces should be designed so that work is conducted at approximately 15% or less of maximum capacity [Chaffin et al. 2006].

Therefore, to minimize the level of effort as a percentage of the maximum capacity, you should help workers use the neutral posture of their joints. However, some joint motion must occur because remaining in a static posture for too long produces several negative consequences and should be avoided. When a worker remains in a static posture, the prolonged application of a load by the muscles can result in fatigue. Also, not moving muscles for a time impedes blood flow, which is needed to bring oxygen and crucial nutrients to the muscles and to remove metabolic waste products. Static postures are avoided when work is dynamic, with the muscles and joints periodically moving. With this in mind, workstations, tasks, and hand tools should be designed to enable workers to use primarily neutral postures and postures that are in relative proximity to the neutral posture. Care should be taken to ensure that awkward postures are not frequent and that high forces are not required while in awkward postures. Figures 1–4 show neutral and awkward postures for the joints (e.g., wrist, elbow, shoulder, and back). These topics will be discussed in more detail in this document.

Special considerations are made for the back and hand. Even though the neutral posture of the back *technically* occurs with the back slightly forward flexed, lifting in a flexed posture can place unwanted forces on the spine itself. Lifting tasks should be performed while the back is not flexed, and the nonflexed posture is often called the neutral posture of the back. The neutral posture for the hand is achieved when the fingers are in a slightly flexed (relaxed) position [Bechtol 1954].

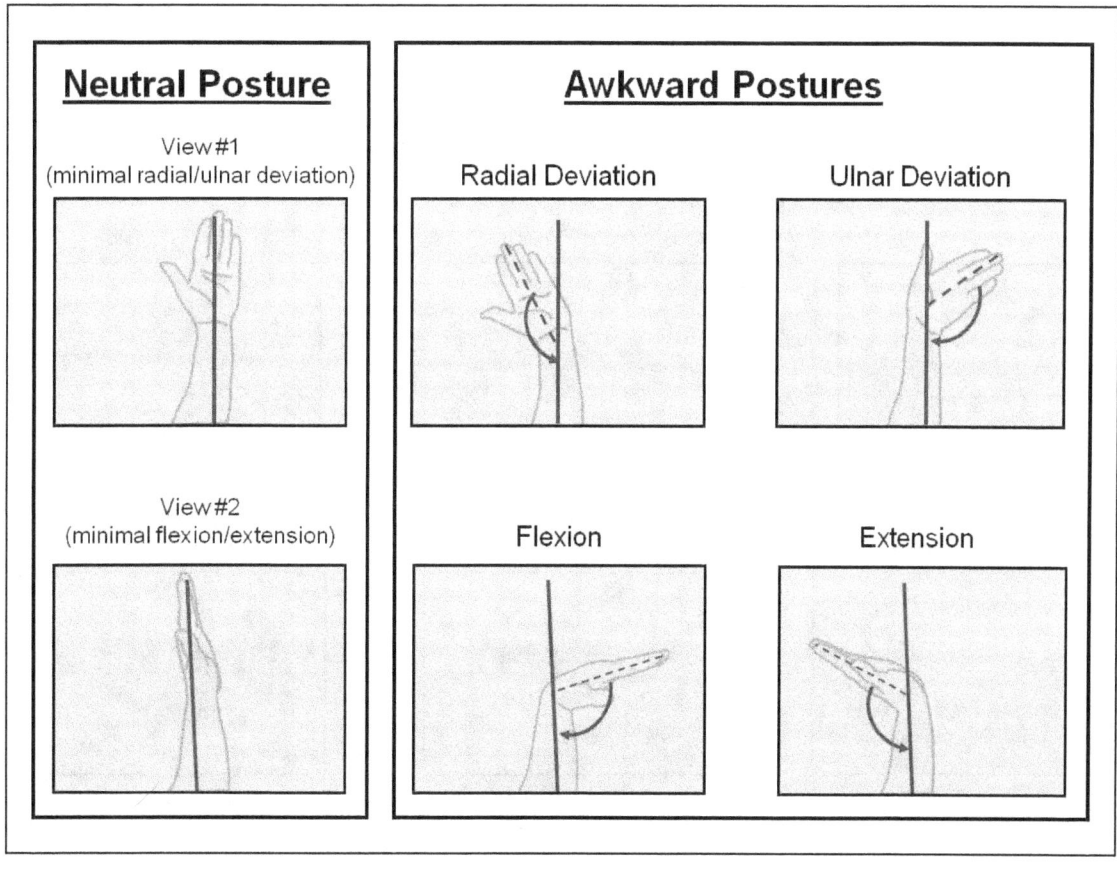

Figure 1. Neutral and awkward wrist postures.

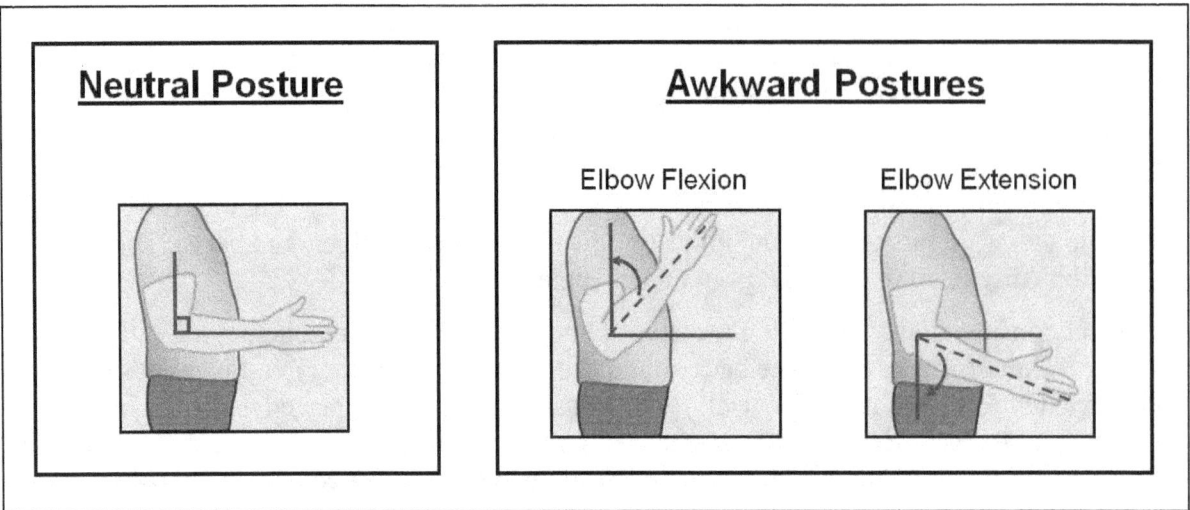

Figure 2. Neutral and awkward elbow postures.

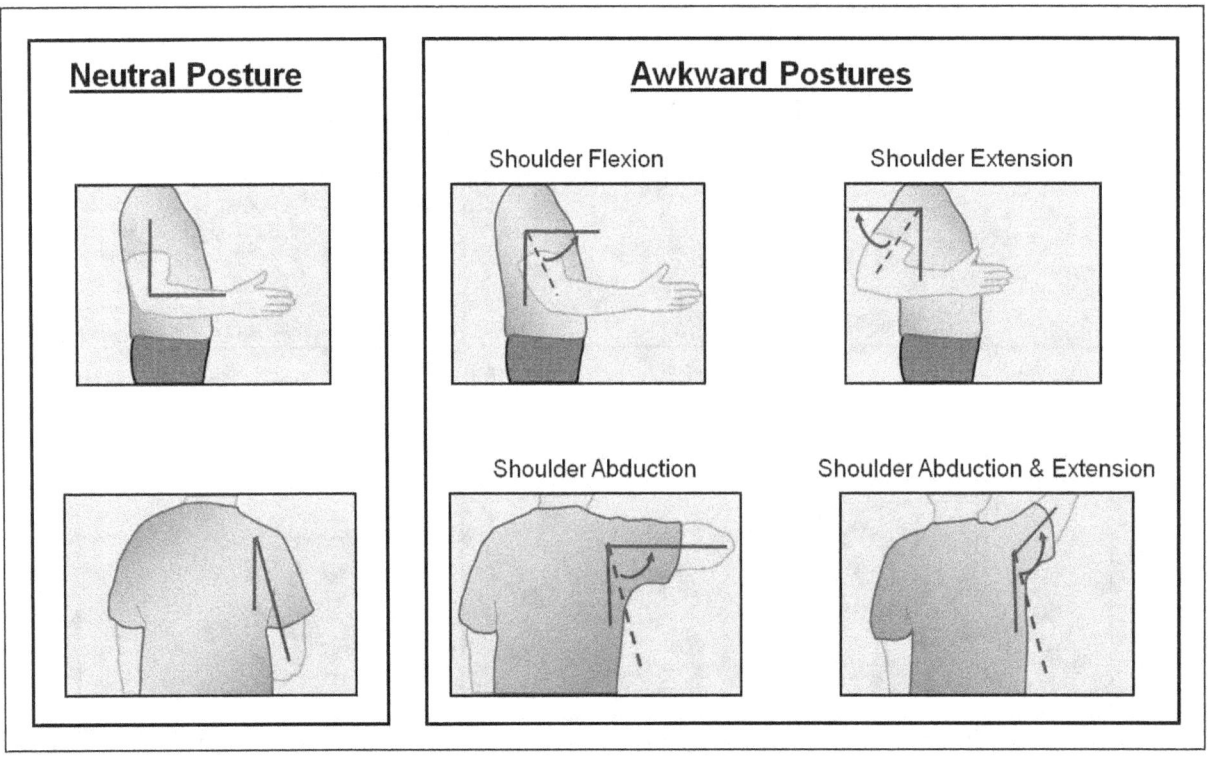

Figure 3. Neutral and awkward shoulder postures.

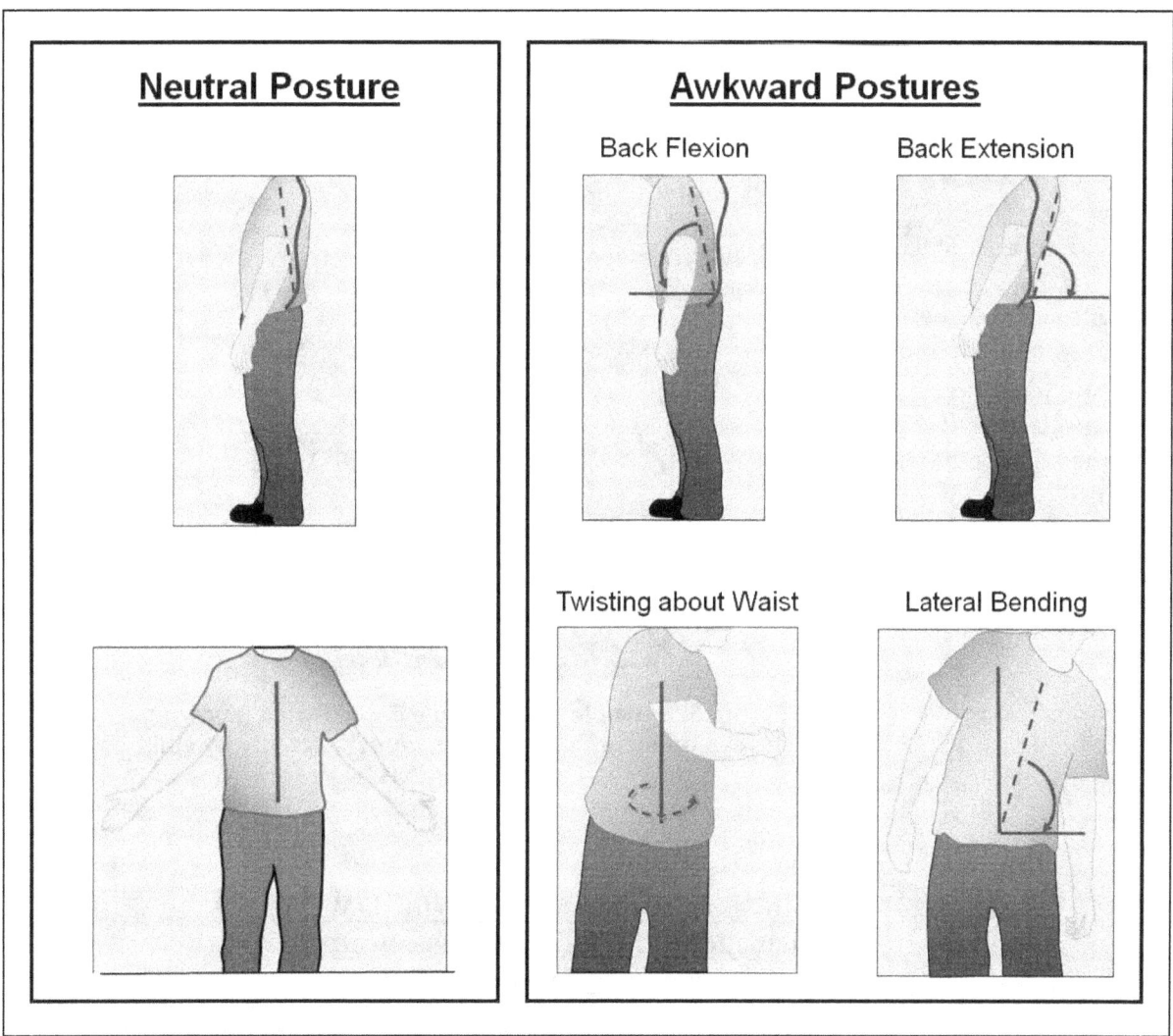

Figure 4. Neutral and awkward back postures.

The following demonstrations are designed to highlight the effect that **awkward** postures have on muscle activity for the wrist, elbow, shoulder, and lower back.

Effects of Postures on Muscle Activity

Objectives

To understand the effect of neutral and awkward postures on muscle activity for the wrist, elbow, shoulder, and lower back

Supplies

Portable EMG device indicating muscle activity via audible sound (Figure 5)

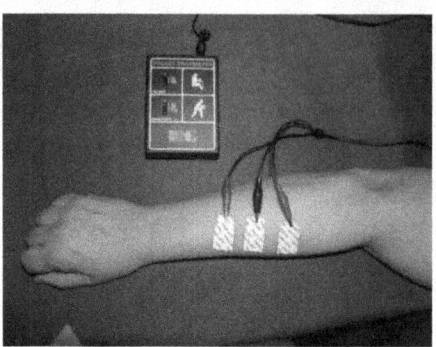

Figure 5. Example of a portable EMG device (showing electrodes on skin) that indicates muscle activity by emitting audible signals.

Step-by-Step Demonstration Method (Wrist). See Introduction video for EMG placement; see WristFlexionExtension video.

1. Place the electrodes on the forearm (see supplemental video clips for more information on electrode placement).

2. Instruct participant to place his or her wrist in the neutral posture and then extend the wrist until it is fully extended and it is clear that the wrist is in an awkward posture (Figure 6).

3. Note that the frequency and volume of the sounds produced by the portable EMG device increase as the joint moves away from the neutral posture and extends into an awkward posture (indicating more muscle activity).

Neutral Compared with Awkward Postures

Effects of Postures on Muscle Activity

Ask the audience to identify tasks at their worksite where they use extended wrist postures or other awkward postures of the wrist (excessive flexion, extension, and radial/ulnar deviations).

Discuss with the audience whether it would be possible to use a neutral posture to perform these tasks. If not, ask the audience why they are limited to awkward postures. Identify changes to workstation design, tasks, tools, or location of materials that may allow neutral postures to be used.

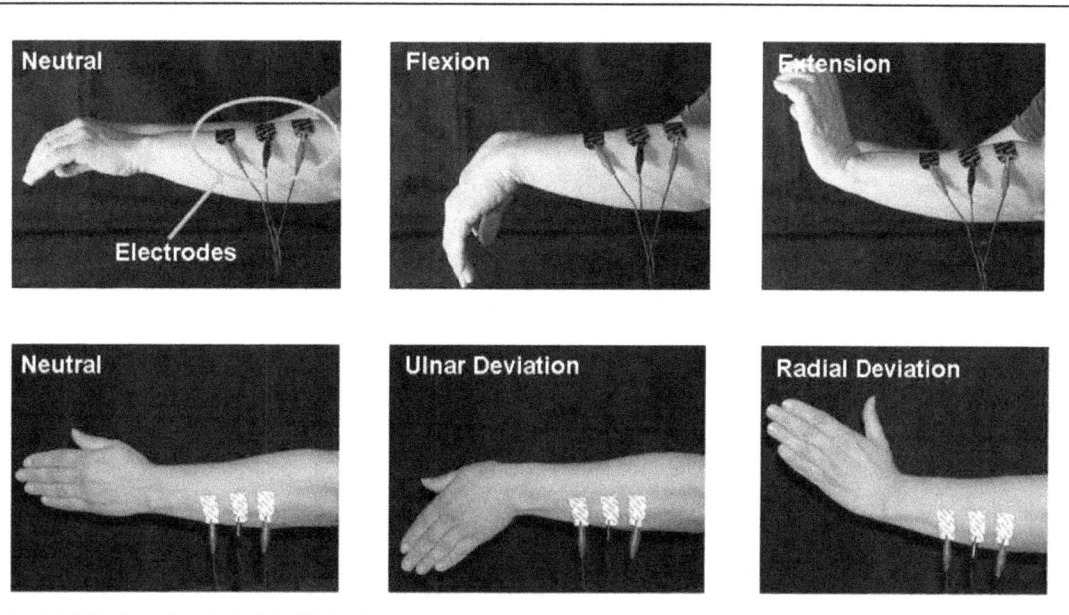

Figure 6. Wrist postures and electrode placement for portable EMG device.

Neutral Compared with Awkward Postures

Effects of Postures on Muscle Activity

NOTE: This demonstration can be used to train workers who use keyboards because it focuses on evaluating wrist posture with the keyboard placed at different positions, including flat, positive, and negative tilt (Figure 7).

Ideal Keyboard Set-up

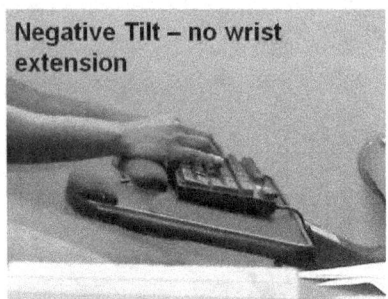

Acceptable (Not Ideal) Keyboard Set-up

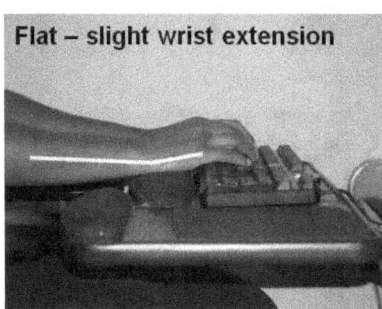

Undesirable Keyboard Set-up

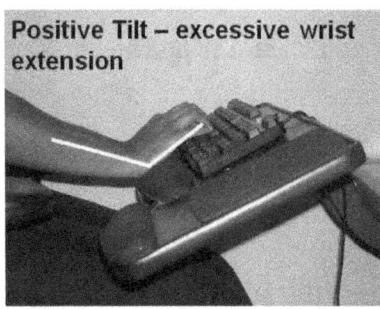

Figure 7. Negative, flat, and positive tilt positions for a keyboard.

Neutral Compared with Awkward Postures

Effects of Postures on Muscle Activity

Step-by-Step Demonstration Method (Elbow). See BicepCurl video.

1. Place the electrodes on the upper arm (Figure 8).

2. Instruct the participant to assume an elbow posture with a 90° angle (neutral) (Figure 8).

3. Note the intensity of the sounds from the portable EMG device.

4. Instruct the participant to raise the forearm so that the elbow angle is less than 90° (flexion). As the elbow angle is decreased, the intensity of the sounds from the portable EMG device will increase (indicating more muscle activity).

5. Ask the audience to identify tasks that require them to use awkward postures for the elbow and how these postures might be avoided.

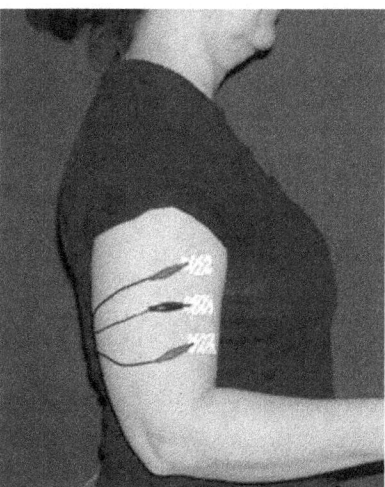

Figure 8. Electrode placement on the upper arm.

Neutral Compared with Awkward Postures

Effects of Postures on Muscle Activity

Step-by-Step Demonstration Method (Shoulder). See ShoulderRaise and ShoulderReach videos.

1. Place the electrodes on the shoulder (Figure 9).

2. Instruct the participant to assume a neutral shoulder posture (Figure 10).

3. Note the intensity of the sounds from the portable EMG device.

4. Instruct the participant to raise his or her arm (abduction) so that it is parallel to the ground. As the shoulder angle is increased, the intensity of the sounds from the portable EMG device will increase (indicating more muscle activity).

5. Instruct the participant to reach above their head as if to change a light bulb. As the shoulder angle is increased further, the intensity of the sounds from the portable EMG device will also increase (indicating more muscle activity).

6. Ask the audience to identify tasks that require them to use awkward postures for the shoulder and how these postures might be avoided.

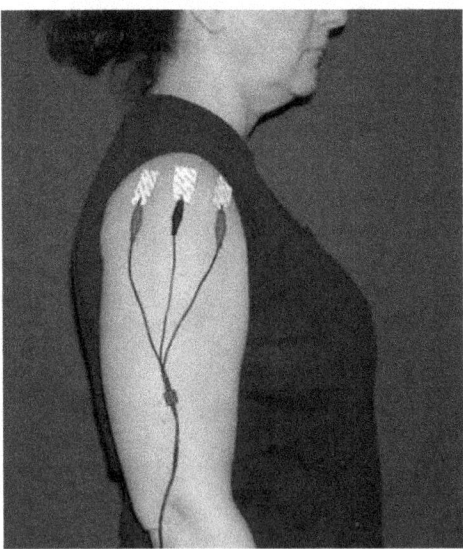

Figure 9. Electrode placement for the shoulder.

Neutral Compared with Awkward Postures

Effects of Postures on Muscle Activity

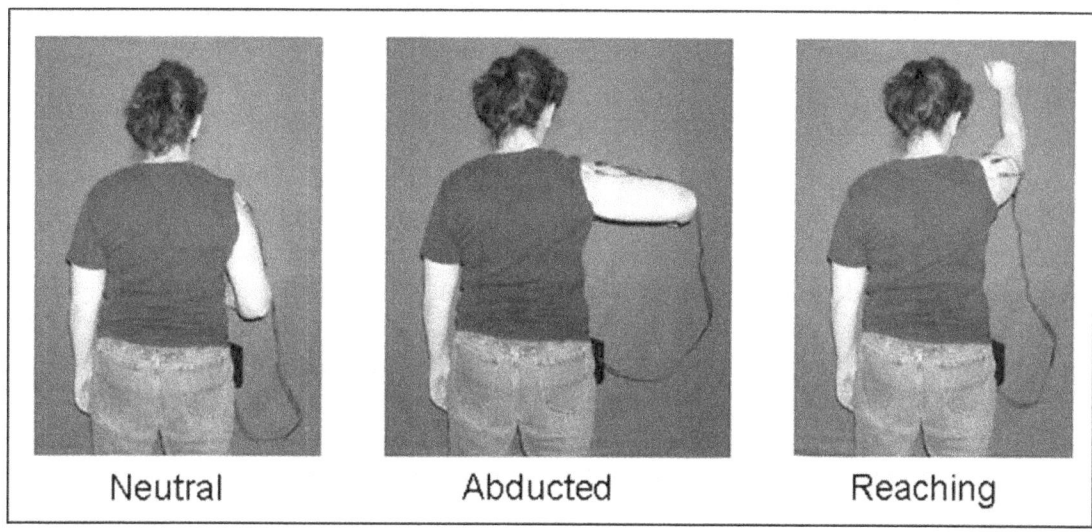

Figure 10. Neutral, abducted, and flexed (reaching) shoulder postures.

Neutral Compared with Awkward Postures

Effects of Postures on Muscle Activity

Step-by-Step Demonstration Method (Low Back). See BackFlexionNoWeight video.

1. Place the electrodes on the low back. (Figure 11).

2. Instruct the participant to slowly lean forward with the back at about a 45°–60° angle, and note the increase in the intensity of the sounds from the portable EMG device (Figure 12). The muscle group being tested is referred to as the erector spinae, which undergoes an eccentric contraction as the trunk flexes forward. The erector spinae helps control the rate at which the torso is lowered by acting against the abdominal muscles that are performing a concentric contraction to flex the trunk. However, make sure the participant does not flex until their torso is fully horizontal since the erector spinae is not as active once the torso comes to rest. The decreased activity, as indicated by a decrease in audible EMG signals, may be confusing to the audience. Before performing this demonstration in front of a group, practice determining the position of the torso when the activity of the erector spinae diminishes. This will help you to advise the participant to avoid going beyond this position. Although muscle activity has decreased at postures near full flexion, the spine continues to be loaded in an undesirable manner.

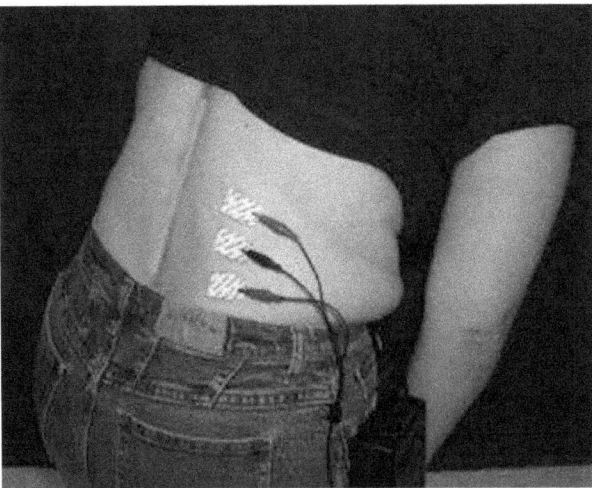

Figure 11. Electrode placement for the back (line indicates location of spine). It is important that the electrodes are placed on the muscles as shown in the photograph. If the electrodes are placed too high on the back, the demonstration will not work properly.

Neutral Compared with Awkward Postures

Effects of Postures on Muscle Activity

3. Ask the participant to slowly return to a standing position. During this motion, the erector spinae are undergoing a concentric contraction in order to raise the weight of the torso upward against the force of gravity. However, when the participant is back in an upright posture, this activity will diminish as will the frequency of the audible signal from the EMG device.

4. Ask the audience to identify tasks at their worksite that require them to work with their backs in flexed postures, and how these postures might be avoided.

Take Home Messages

Whenever possible, workers should work in a neutral posture to reduce muscle activity and reduce fatigue.

Neutral

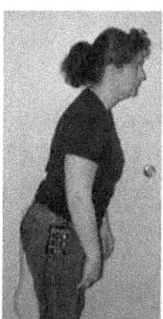
Moderately Flexed

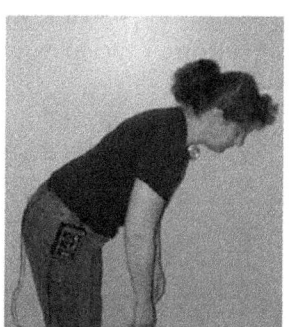
Highly Flexed

Figure 12. Neutral, moderately flexed, and highly flexed postures of the back.

Neutral Compared with Awkward Postures

Wrist Angle and Grip Strength

Objectives

To increase awareness of how posture affects force production, capabilities, and worker fatigue

To discuss tasks that require workers to use awkward postures while exerting force

Supplies

Hand dynamometer (**Figure 13**)

Stopwatch

Step-by-Step Demonstration Method. See PowerGrip video.

1. Place the hand dynamometer in the participant's hand and make sure his/her hand is in the neutral posture as shown in Figure 13.

2. Instruct the participant to squeeze with maximum force, and record the force he/she was able to produce.

3. Instruct the participant to rotate his or her wrist into a position of radial deviation (awkward posture).

4. Instruct the participant to squeeze with maximum force, and record the force he or she was able to produce. The force should be less than the force obtained with the wrist in the neutral posture.

5. Instruct the participant to rotate his or her wrist into a position of ulnar deviation (awkward posture, see Figure 13). Some people may not have sufficient range of motion to move their wrist into this posture. Before performing this demonstration, determine if the participant selected for the demonstration can achieve this posture.

6. Instruct the participant to squeeze with maximum force, and record the force he/she was able to produce. The force should be less than that produced with the wrist in the neutral posture.

7. Ask a few more audience members to participate. The trend will be the same for all participants even though the maximum forces they can produce will vary.

Neutral Compared with Awkward Postures

Wrist Angle and Grip Strength

8. Now ask each participant to exert a force of 20 lbs while in the neutral posture, and hold that force as long as possible. Record the length of time the participant maintains the posture. Then, ask the participants to do the same using wrist positions with radial and ulnar deviations; because fatigue occurs sooner in these postures, the length of time the force can be maintained for these postures should be less than the time the force can be maintained using a neutral posture.

9. Ask the audience to identify tasks at their worksite where awkward postures of the wrist occur.

NOTE: Dramatic results will also be seen if this demonstration is performed with the wrist in extension or flexion as shown in Figure 1.

Neutral Compared with Awkward Postures

Wrist Angle and Grip Strength

Take Home Messages

When body joints are in awkward postures, maximum force produced decreases.

Muscle fatigue will occur earlier when working in an awkward posture instead of a neutral posture.

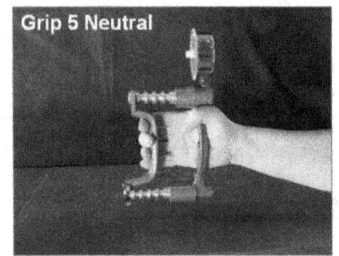
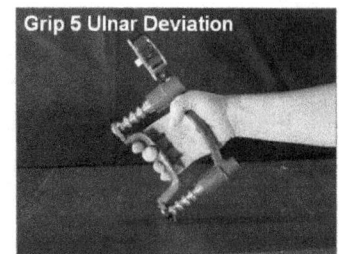
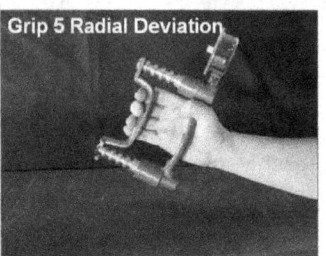
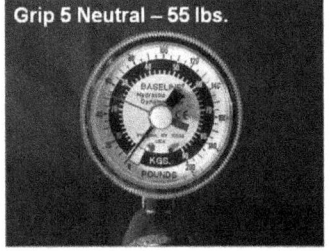
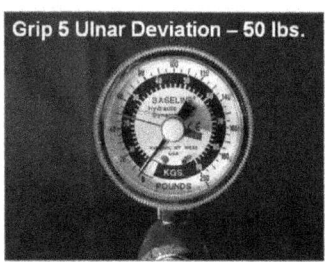
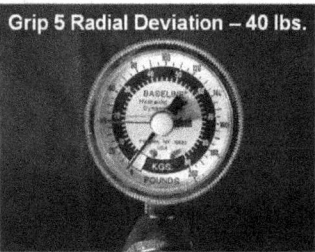

Figure 13. Hand dynamometer showing how wrist angle affects force production for neutral, ulnar deviation, and radial deviation wrist postures.

Neutral Compared with Awkward Postures

SECTION 3: GRIP TYPES

Principles

- **Force generated with a pinch grip is about 15%–25% of force generated with a power grip.**

- **Use a power grip when higher forces are required.**

- **Use a pinch grip when precise movements are needed, and the force required is low (< 2 lbs).**

- **Research shows the design width of power grips should be 1.75 to 3.75 inches.**

In general, an object can be grasped using one of two methods: a pinch grip or a power grip (Figure 14). A power grip curls the fingers toward the palm; a pinch grip presses the thumb against the fingers of the hand or an object, and does not involve the palm. The amount of force that can be generated depends on the type of grip and the width of the grip.

Three types of pinch grips can be used:

Tip pinch—using only the tips of the fingers and thumb (holding a bead)

Chuck pinch—using the thumb and first two fingers (holding a pencil)

Lateral pinch—using the thumb and side of the first finger (holding a key)

For a given force, using a pinch grip is biomechanically more stressful than using a power grip. The amount of force one is capable of exerting is greater for the power grip than for the pinch grip. A general rule of thumb is that the force generated with a pinch grip is about 15%–25% of the force generated with a power grip, depending on the type of pinch grip and the worker's individual force capability. The amount of force exerted also varies among the three types of pinch grips. When using a tip pinch, the force exerted is 71%–72% of the lateral pinch force; when using a chuck pinch, the force exerted is 98% of the lateral pinch force. The amount of force generated by power grips and pinch grips also varies depending on the width of the grip. For a power grip, the maximum force is generated with a grip width of 1.75–3.75 inches. For a pinch grip (type not specified by citation), the maximum force is generated with a grip width of 1–3 inches [Chengalur et al. 2004].

A pinch grip provides more control because the thumb joint is highly movable and precise. In contrast, minimal control is associated with the power grip as the fingers move as one entity and only in one direction (flexion). For these reasons, pinch grips are typically used for short-duration, low-force, and precision tasks because they require minimal force exertion but high control (e.g., tightening or removing eyeglass screws). In general, tasks that are done repeatedly and require 2 lbs or more of force should not involve pinch grips. For example, tasks that require using a power drill are ideally suited to the use of a power grip because the neutral posture for the fingers is a slightly flexed position [NIOSH 2004].

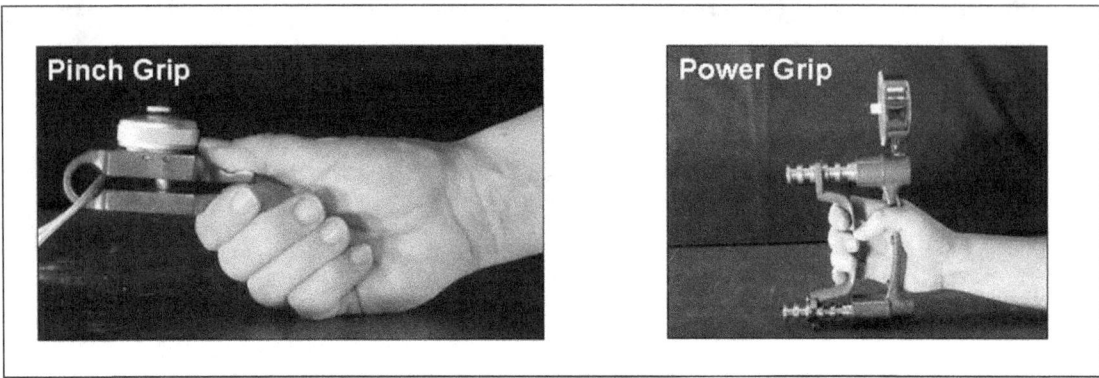

Figure 14. Pinch (lateral) grip and power grip.

NOTE: Grip type is greatly influenced by hand-tool design. Therefore, a separate section in this document (see Section 4) has been devoted to hand-tool selection and use, and follows this section on grip types.

Power Grip Compared with Pinch Grip

Objectives

To increase awareness that:

Maximum force generated using a power grip is greater than when using a pinch grip.

Force production capabilities differ among individuals for both pinch and power grips.

Pinch grips should be avoided when possible because it places high demands on the hand and produces less force than a power grip.

Supplies

Hand dynamometer that measures pinch-grip strength (Figure 15)

Hand dynamometer that measures power-grip strength (Figure 15)

Step-by-Step Demonstration Method. See LateralPinchGrip, TipPinchGrip, ChuckPinchGrip, and PowerGrip videos.

1. Place dynamometer for pinch grip measurements between the participant's thumb and forefinger as shown in Figure 15 (lateral).

2. Instruct the participant to exert maximum force.

3. Note the maximum force produced.

4. Place the dynamometer for power-grip measurements in the palm of participant's hand, as shown in Figure 15.

5. Instruct the participant to exert maximum force.

6. Note the maximum force produced.

7. Ask several participants to exert maximum forces using both pinch (lateral, tip, and chuck) and power grips. Record these results.

8. Discuss the differences between the maximum forces generated for the different types of grips.

9. Discuss the differences among individuals in generating forces using all types of grips.

Grip Types

Power Grip Compared with Pinch Grip

10. Discuss specific tasks that would be appropriate for using either type of grip.

11. Identify specific tasks the worker performs that uses the different types of grips, and determine if the grip is appropriate for the task requirements.

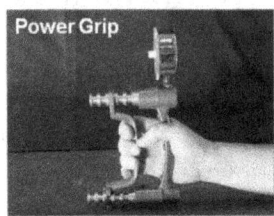

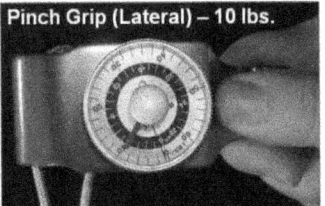 

Figure 15. Example of the maximum forces generated for a pinch grip (lateral) and a power grip.

Take Home Messages

A pinch grip should be used only for precision tasks that require minimal forces to be generated.

In general, pinch grips should be avoided for any length of time, regardless of the force required.

A power grip should be used for tasks that require larger forces, that do not require high degrees of precision and dexterity.

Grip Types

Pinch Grip Strength and Applications

Objectives

To increase awareness of individual maximum capabilities for a pinch grip

To discuss tasks that are performed with a pinch grip

To understand force and repetition requirements of the task

Supplies

Hand dynamometers that measures pinch grip strength (Figure 16)

Step-by-Step Demonstration Method

1. Place hand dynamometer between participant's thumb and forefinger (lateral pinch grip) as shown in Figure 16.

2. Instruct the participant to exert maximum force.

3. Note the maximum force produced.

4. Ask several other audience members of different sizes (i.e., weight, height) and/or gender to perform steps 1–3 and compare their maximum forces; this demonstrates the effects of anthropometry (i.e., size variability among people). Encouraging audience members to compete for the largest force production often increases participation and friendly competition. Make sure to point out to the audience how much the forces produced varied across the group.

5. Instruct the participants to exert 2 lbs of force to provide them with a general understanding of what it feels like to exert that level of force.

6. Ask the audience to identify tasks at their worksite where they use a pinch grip and exert more than 2 lbs of force. Repetitive tasks for which a pinch grip is used should also be avoided. Ask the audience to identify repetitive tasks at their worksite for which they use a pinch grip.

7. Discuss the possibility of using a better tool or workstation design to avoid using pinch grips.

Grip Types

Pinch Grip Strength and Applications

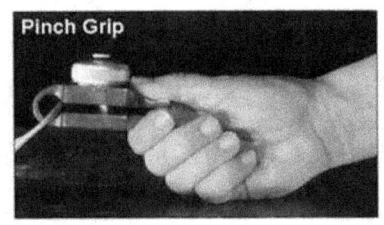 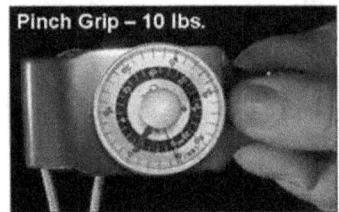

Figure 16. Example of a pinch grip (lateral) and the resulting maximum force.

Take Home Messages

Maximum forces exerted with a pinch grip vary among workers.

A pinch grip should not be used when high forces or repetition are required.

Pinch grips should be used only for tasks that require small forces (< 2 lbs).

Grip Types

Power Grip: Effect of Grip Width

Objectives

To increase awareness of how grip width affects maximum strength for a power grip

To increase awareness of individual maximum capabilities for a power grip

To discuss whether or not the grip width necessary to perform tasks is appropriate

Supplies

Hand dynamometer that allows grip strength to be evaluated for multiple grip widths (Figure 17)

Step-by-Step Demonstration Method. See PowerGrip, NarrowGrip, and WideGrip videos.

1. Place the hand dynamometer in the participant's hand, and instruct him or her to place the wrist in a neutral posture (Figure 17).

2. Measure the maximum force the participant can produce using three to five different grip widths. If using only three different grip widths, use grips 1, 3, and 5 as shown in Figure 17.

3. Record the force produced for each grip width and compare these values across the different grip widths. You should notice that for very wide grips and for very narrow grips, the participant will not be able to produce as much force as with the intermediate grips (Figure 18).

4. Ask several other audience members of different sizes (i.e., height, weight) and/or gender to perform steps 1–3, and compare their maximum forces; the forces produced will vary, showing the effect of anthropometry). However, the maximum force for each participant should be produced for a grip width of 1.75 to 3.75 inches.

5. Ask the audience to identify tasks at their worksite where a power grip is required at or near their minimum or maximum grip width capacity. Determine whether or not these tasks require workers to exert forces near their maximum capabilities.

6. Discuss with the audience how they may use a better tool or workstation design to avoid using power grip widths that are too narrow or too wide.

Grip Types

Power Grip: Effect of Grip Width

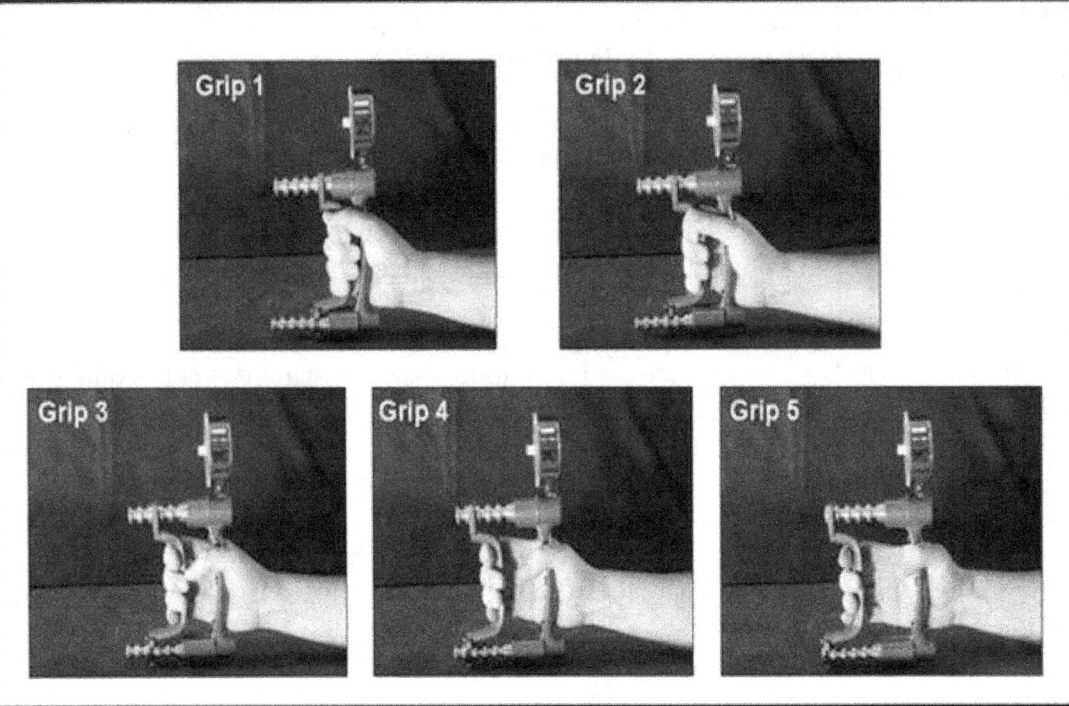

Figure 17. The power grip is shown for five different grip widths. The narrowest grip is Grip 1; the width increases for each subsequent grip, with Grip 5 being the widest grip.

Grip Types

Power Grip: Effect of Grip Width

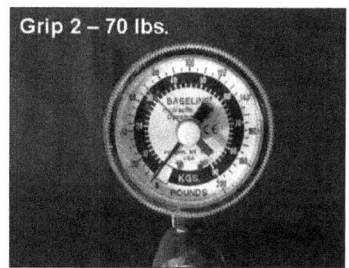

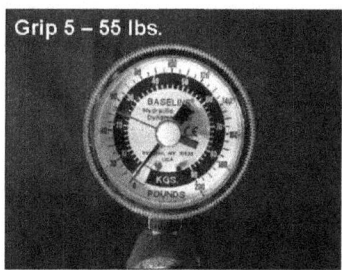

Figure 18. Maximum–force output for each grip width. Note that, for this participant, Grip 2 had the highest force production.

Take Home Messages

Maximum force produced with a power grip varies with grip width.

Maximum forces exerted for a power grip vary among workers.

Tools and workstations should be designed so that workers may use optimum power grip widths (1.75–3.75 inches).

Grip Types

SECTION 4: HAND-TOOL SELECTION AND USE

Principles

- Select tools that allow neutral postures to be used.

- Use tools with handles designed for a power grip.

- Use tools with handles that are appropriately sized and shaped for the user's hand.

- Use tools with built-in features (e.g., springs that open tool handles) that minimize forceful exertions required to use the tool.

- When operating heavy tools, ensure they accommodate using both hands to support the tool's weight.

Hand-tool design can play an important role in the reduction of MSDs. A tool that is designed with consideration for the worker's tasks can greatly reduce the worker's exposure to risk factors for MSD. However, using a poorly designed tool or an inappropriate tool negatively impacts the entire body by dictating the postures assumed by the worker to complete the task, and increasing the resulting forces exerted by the worker. Such tools can also directly apply unwanted forces or vibrations to other body parts.

Several factors should be considered when purchasing or selecting a hand tool. The topics discussed in the above sections all play a role in whether or not a hand tool is designed with the worker or task in mind. Some of these points will become clearer after performing the demonstrations in this section. Before performing these demonstrations, consider the following questions related to the safety of the tools you use:

- Does the orientation of the handle allow the worker to use neutral joint postures?
- Does the size of the handle allow for the midrange of grip (1.75–3.75 inches; hand in the shape of a "C") width when using a power grip?
- Does the handle extend past the palm?
- Is the handle shape contoured to fit the palm?
- If a pinch grip is required, is the force the worker must exert < 2 lbs.?
- For heavier tools, such as power tools, do the features of the tool allow the worker to support the tool's weight with both hands?

To obtain a detailed checklist for hand-tool selection, refer to the NIOSH publication "Easy Ergonomics: A Guide to Selecting Non-Powered Hand Tools" [NIOSH 2004].

Tool-Handle Size and Shape

Objectives

To understand why handle size is important when selecting tools

Supplies

One screwdriver, whose handle has a diameter size that complements the hand and a comfortable, appropriate shape

One screwdriver, whose handle has a smaller diameter size that does *not* complement the hand and does *not* have a comfortable, appropriate shape

Wood block with screw

Clamp to affix wood block to tabletop

Step-by-Step Demonstration Method

1. Clamp wood block and screw to tabletop.

2. Instruct the participant to grasp the appropriately sized screwdriver that has the larger diameter handle. Instruct the participant to show the audience how he/she is gripping the screwdriver (Figure 19).

3. Repeat the previous step using the other screwdriver, with a smaller diameter handle that does not complement the hand. Discuss observed differences in gripping the two screwdrivers.

4. Instruct the participant to drive a screw with both screwdrivers. Ask them which one feels more comfortable in his or her hand and is easier to grasp when the screw starts to provide resistance. The handle with the larger diameter that complements the hand should make it easier for the participant to apply torque to drive the screw.

5. Discuss with the audience the fact that selecting a hand tool with a handle that complements the hand reduces the effort needed to accomplish the task, thus reducing fatigue and the required muscle activity; this, in turn, reduces discomfort while using the tool.

Hand-Tool Selection and Use

Tool-Handle Size and Shape

Take Home Messages

Tools whose handles are sized and shaped to complement the hand, require less effort to use, thereby reducing the muscle fatigue that leads to discomfort.

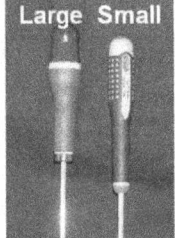

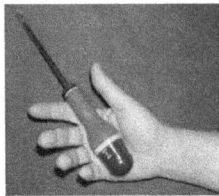
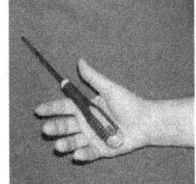

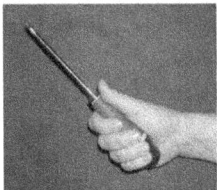
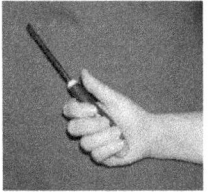

Figure 19. Evaluating the effect of tool-handle diameter.

Hand-Tool Selection and Use

Tool-Handle Orientation

Objectives

To demonstrate how tool-handle adjustability or orientation may allow for neutral postures to be adopted

To encourage workers using and purchasing tools to consider how the tool will be applied and whether or not a different tool, or tool configuration, would be more appropriate

Supplies

Screwdriver (battery-powered) with pistol-grip and inline-grip capabilities (Figure 20)

Wood block with screw in block

Clamp to hold wood block in place while screw is being driven with screwdriver

Step-by-Step Demonstration Method

1. Clamp the wood block to a tabletop so that the block is perpendicular to the table and the screw is driven parallel to the tabletop.

2. Place the screwdriver in the inline position, and instruct the participant to begin driving the screw.

3. Place the screwdriver in the pistol-grip position, and instruct the participant to begin driving the screw.

4. Ask the participant if he or she can feel a difference between the two techniques. Ask the audience which grip would be best to use for this task (answer: pistol grip).

5. Clamp the wood block to a tabletop so that the block is parallel to the table and the screw is driven perpendicular to the tabletop.

6. Place the screwdriver in the inline position, and instruct the participant to begin driving the screw.

7. Place the screwdriver in a pistol-grip position, and instruct the participant to begin driving the screw.

8. Ask the participant if he or she feels a difference between the two techniques. Ask the audience which grip would be best to use for this task (answer could vary depending on height of participant relative to the tabletop which affects his/her wrist and shoulder angle).

Hand-Tool Selection and Use

Tool-Handle Size and Shape

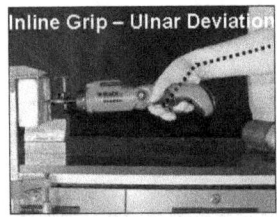 Bent Wrist → Poor Wrist Position

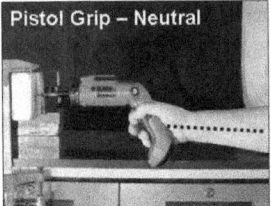 Straight Wrist → Ideal Wrist Position

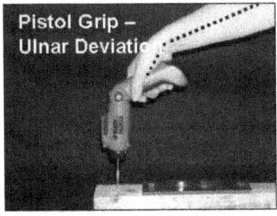 Bent Wrist → Poor Wrist Position

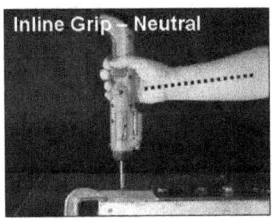 Straight Wrist → Ideal Wrist Position

Figure 20. Examples of situations in which a pistol grip and inline grip would be useful as a means for keeping the wrist in a neutral posture.

Note: This demonstration can be done without actually driving a screw into a wood block. You can ask the participant to simulate driving a screw into a tabletop and into the wall. The differences in postures can be observed and discussed by the audience.

Take Home Messages

Adjustability in tools, or multiple tool designs, is important because it allows for neutral postures to be adopted

When selecting or purchasing a tool, consider the ability of the tool's handle to be adjusted in multiple positions to keep the wrist in a neutral posture.

Hand-Tool Selection and Use

Features to Reduce Forceful Exertions

Objectives

To increase awareness that some tools have design features that reduce forceful exertions when the tool is used to perform a task

Supplies

Spring-loaded, needle-nose pliers (Figure 21)

Needle-nose pliers that are not spring loaded (Figure 21)

Portable EMG device with audible sounds to indicate muscle activity (optional) (Figure 5)

Step-by-Step Demonstration Method

1. If using a portable EMG device, place the electrodes on the forearm as shown in Figure 5.

2. Adjust the output sounds to a range where minimal sounds are heard when the participant wiggles his or her fingers.

3. Instruct the participant to close and then open the spring-loaded, needle-nose pliers.

4. Note the intensity of the sounds from the portable EMG device.

5. Instruct the participant to close and then open the needle-nose pliers that are not spring loaded.

6. The intensity of the sounds from the portable EMG device will be increased when using the nonspring-loaded pliers compared to the spring-loaded pliers. When using pliers for a work activity, individuals often use their dominant hand to both open and close the pliers. Spring-loaded pliers remove the need to open the pliers and reduces the force requirements on the hand.

7. Ask the participant if he or she can feel a difference between the two tools.

Hand Tool Selection and Use

Features to Reduce Forceful Exertions

8. Discuss with the audience that, even though a tool may be spring loaded, it may be still difficult to use if the resting-grip width is large (see Section 3 concerning grip widths).

9. Discuss with the audience other design features that reduce forceful exertions, such as counterbalances, ratcheting tools, keyless drill chucks.

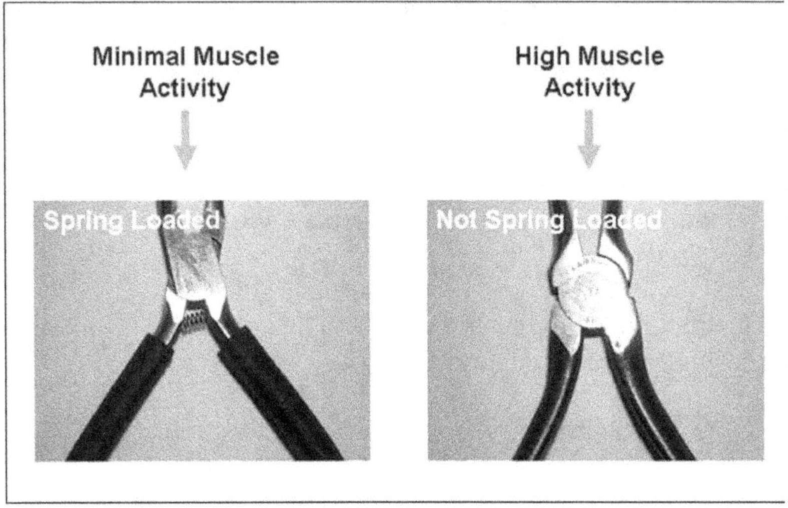

Figure 21. Two types of pliers, one with a spring that reduces forceful exertion when opening the jaw, and one without a spring.

NOTE: Another example of reduced forceful exertions as a result of design features is the insertion of a bit into a screwdriver that requires manual tightening of the chuck with a key, as compared to a screwdriver with a chuck that is simply pushed down and then released.

Take Home Messages

Select tools with features that reduce forceful exertions when preparing the tool for use and operating the tool during the task.

Hand-Tool Selection and Use

One- and Two-Handed Tools

Objectives

To demonstrate that muscle activity decreases when forces are distributed across both arms instead of just one arm

To inform workers that they should purchase and use tools with appropriate power because too much power for the job may result in difficulty controlling the tool, increased fatigue, and poor-quality workmanship

Supplies

Electric drill with capability to hold drill with one hand or with two hands using an additional handle (Figure 22)

Wood block with a screw

Clamp to affix wood block to a table

Step-by-Step Demonstration Method

1. Instruct the participant to hold the drill at waist height and then at shoulder height with one hand, and note the degree of effort required to hold the drill for both.

2. Instruct the participant to grab the additional handle with the second hand, and again hold the drill at waist and then shoulder height. It should be easier with both hands on the tool because the muscles from both arms are now being used to hold the drill; this reduces the force produced by each individual muscle.

3. Explain to the audience that, because force in each muscle has decreased, fatigue will set in later than if the task was performed with only one hand.

4. Clamp down the wood block with the screw.

5. Instruct the participant to drive the screw while holding the drill with only one hand.

6. Instruct the participant to drive the screw while holding the drill with both hands.

7. Ask the participant if he or she feels more control when using two hands.

Hand-Tool Selection and Use

Features to Reduce Forceful Exertions

8. Ask the audience if they perform tasks that should be performed with a tool that has the one- and two-handed design feature.

8. Discuss the concept relating to the tool's power and the worker's ability to control the tool—as the power of the tool increases, a worker's ability to control the tool decreases; this may result in poor-quality workmanship. Also, buying tools with excessive power may have negative consequences for the worker by requiring greater muscle exertions and causing an earlier onset of fatigue.

Figure 22. Example of one-handed and two-handed drilling.

Take Home Messages

Select tools that are properly sized in overall dimensions, weight, and power for the specific task. Too much weight and power can increase fatigue in the worker, and result in poor-quality workmanship.

When purchasing heavy power tools, consider features that allow the tool to be held with both hands.

When operating heavy tools, take advantage of features that allow for greater control of the tool and less fatigue.

Hand-Tool Selection and Use

SECTION 5: FATIGUE FAILURE AND BACK PAIN

Principles

- **Repeated lifting, even at submaximal levels, may eventually lead to damage of the spine (fatigue failure).**

- **Substantially reducing loads placed on the spine can greatly minimize the risk of fatigue failure.**

The spine consists of a column of bones called vertebrae that are separated by flexible discs (Figure 23). The discs serve as cushions and allow the spine to assume many postures. Degeneration of these discs is a common source of back pain, which is thought to result from a loss of disc nutrition [Adams et al. 2006]. Because discs do not have a blood supply, they rely on obtaining their nutrition from the adjacent bones (or vertebrae). Normally, nutrients flow from the vertebrae to the disc through structures called vertebral endplates. The endplates are on the top and bottom of each vertebrate. Unfortunately, these endplates may fracture if an excessive force or repeated loads are placed on them by the contracting back muscles, as occurs when lifting [Brinckmann et al. 1989].

Researchers believe that endplate fractures usually occur through repeated loading, by a process known as fatigue failure [Bogduk 1997; Brinckmann et al. 1988; Adams et al. 1995, 2006; Gallagher et al. 2005; Marras 2008]. Fatigue failure begins when a load causes a small crack in a vertebral endplate. Subsequent loads (e.g., repeated lifting) will cause this crack to expand, leading to a large fracture [Brinckmann et al. 1988]. The body heals this fracture with scar tissue, but the scar tissue does not allow nutrients to get to the disc, causing it to degenerate [Bogduk 1997]. As the disc degenerates, fissures and tears in the disc will begin to appear. When these fissures or tears extend to (or occur in) the outer portions of the disc, a painful inflammation may occur. Unfortunately, because the disc has a decreased blood supply, repair of the tissues is a slow process [Bogduk 1997]. At the same time, the disc often continues to become loaded during activities of daily living, which may result in additional damage to the disc, even while repairs are ongoing. The slow healing, combined with continuous loading and trauma to the tissues, is thought to lead to a vicious cycle of chronic pain and inflammation [Barr and Barbe 2004].

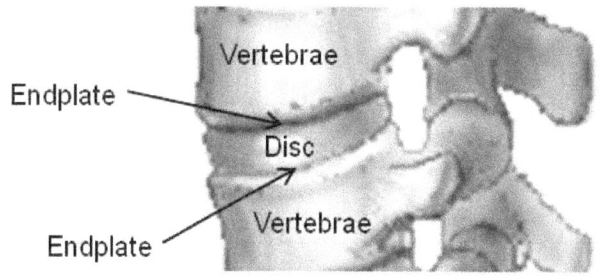

Figure 23. Image of vertebrae, disc, and endplates.

Fatigue Failure

Objectives

To introduce workers to the concept of fatigue failure

To reinforce the importance of minimizing object weight, lever arm, barriers, and repetition of manual lifting tasks

Supplies

One pen cap

One paper clip for each audience member

Step-by-Step Demonstration Method

1. Take intact pen cap and bend "tail" once.

2. Show audience members that there is a discoloration at the spot where the bending occurred, which is a visual example of subfailures occurring.

3. Continue to bend the pen cap about five times, and then show the audience that the discoloration has expanded.

4. Explain to the audience that, if you continue to bend the pen cap, it would eventually fail.

5. Explain to the audience that, for some materials (e.g. paper clip, vertebrae), fatigue failure is not visible.

6. Distribute one paper clip to each audience member.

7. Ask the audience to bend the paper clip back and forth, and count the number of cycles it can withstand before breaking.

8. Ask various audience members how many cycles it took before the paper clip failed; emphasize that the number of cycles varies for the paper clips as no one paper clip is exactly the same as another. This is also true for people and their vertebrae. Just like with the paper clips, some worker's will experience fractures in their vertebrae very quickly as others require many cycles despite undergoing the same loading conditions.

9. Show the graph in Figure 24 to the audience.

Fatigue Failure and Back Pain

Fatigue Failure

10. Explain that every type of material has an ultimate load (i.e., the load at which it fails when that load is applied only once). The graph in Figure 24 illustrates the amount of loads the spine can handle without breaking. Because every spine is unique, the ultimate load varies somewhat for each spine. Thus, the y-axis of this graph represents the percentage of ultimate load. The x-axis represents the number of cycles a load was applied. For example, if you applied a load to a spine that was 80% of its ultimate load, you would be able to apply that load 100 times before the spine would fail. Likewise, if you applied a load that was only 50% of its ultimate load, you could apply that load 1,000 times before failure. If you applied a load that was only 30% of its ultimate load, you could complete an infinite number of loading cycles without the spine ever failing.

11. Explain to the audience that this means they are not "doomed to having a back injury." Rather, if the load applied to the spine is decreased substantially, they could perform their job an infinite number of times and never injure their spine. You may also increase your core strength to better handle loads—a balanced body in terms of abdominal and back strength makes a more stable core when trained together.

12. Discuss ways to reduce the load applied to the spine, such as decreasing the weight of the object, reducing the moment arm (see Section 6), removing barriers, and eliminating twisting and back flexion.

Fatigue Failure and Back Pain

Fatigue Failure

Take Home Messages

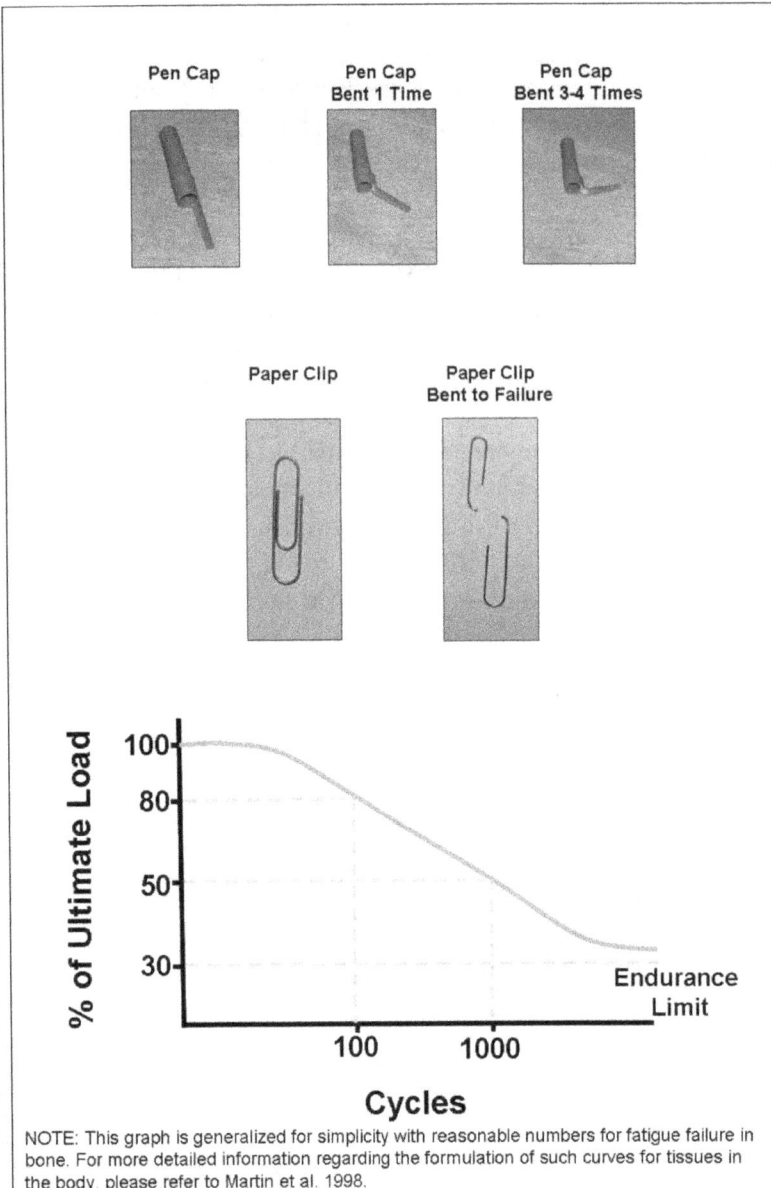

Often, the vertebrae of the back can have multiple subfailures that are not visible but can result in complete failure over time.

The number of cycles that lead to failure of the vertebrae varies across the population.

Efforts should be made to substantially decrease loading of the spine.

NOTE: This graph is generalized for simplicity with reasonable numbers for fatigue failure in bone. For more detailed information regarding the formulation of such curves for tissues in the body, please refer to Martin et al. 1998.

Figure 24. A pen cap that is bent multiple times visually shows fatigue; a paper clip shows the result of failure. The graph (generalized for bone) illustrates how the same load, lifted many times, may ultimately, over time, lead to failure.

Fatigue Failure and Back Pain

SECTION 6: MOMENT ARMS AND LIFTING

Principles

- **Reduce the weight of the object being lifted.**

- **Keep loads close to the body when lifting.**

The best way to prevent low-back pain is to prevent the initial fatigue failure of the vertebral endplates. In general, for a given task, if the forces exerted by back muscles are high (e.g., in heavy lifting), fatigue failure will occur more quickly. However, if forces produced by the low-back muscles are decreased, the risk of injury also decreases [Brinckmann et al. 1988].

Forces produced by the lower back muscles can be reduced by minimizing the weight being lifted or carried. However, those forces can also be reduced by minimizing the moment (see Glossary) or by minimizing the moment arm (i.e., lever arm). When lifting an object, as shown in Figure 25, the moment arm is the horizontal distance between the object and the person. As this distance increases, the moment (i.e., torque), involving the worker's back also increases. The muscles of the lower back must produce more force to counteract this moment so that the person does not fall forward. Even light objects can cause large forces in the lower back if those objects are lifted or carried farther away from the body.

Weight and moment arm are not the only considerations in determining forces produced by the lower back muscles. Other factors, mostly related to the object being lifted, should also be briefly mentioned. The size and shape of the object and the handholds on the object affect the worker's lifting style. Also, the existence of physical barriers that separate the worker from the object to be lifted plays a role in the forces exerted in lifting the object because barriers force a worker to hold an object farther away from his or her body while the worker moves the object over the barrier. A barrier often requires the worker to lift or hold an object incorrectly. The distribution of the weight across the object itself is also a consideration because an awkward weight distribution can also cause the worker to lift and carry the object incorrectly. Detailed information about these factors can be found in Waters et al. [1993] and [NIOSH 1994].

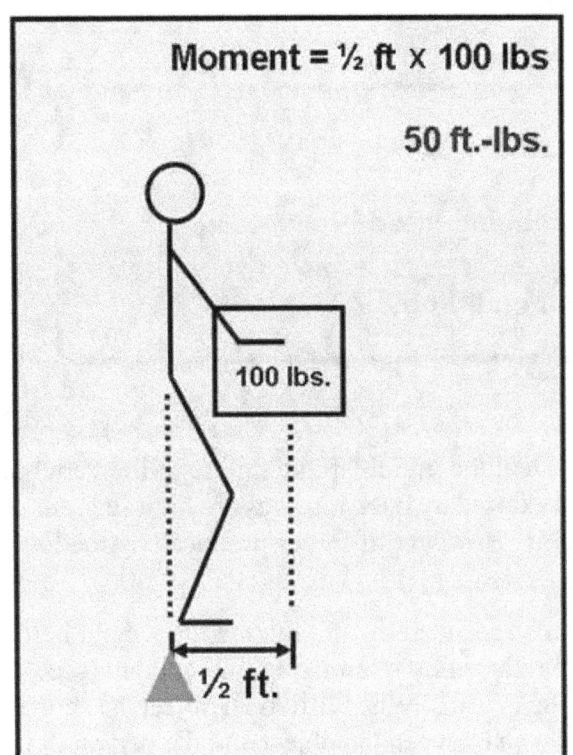 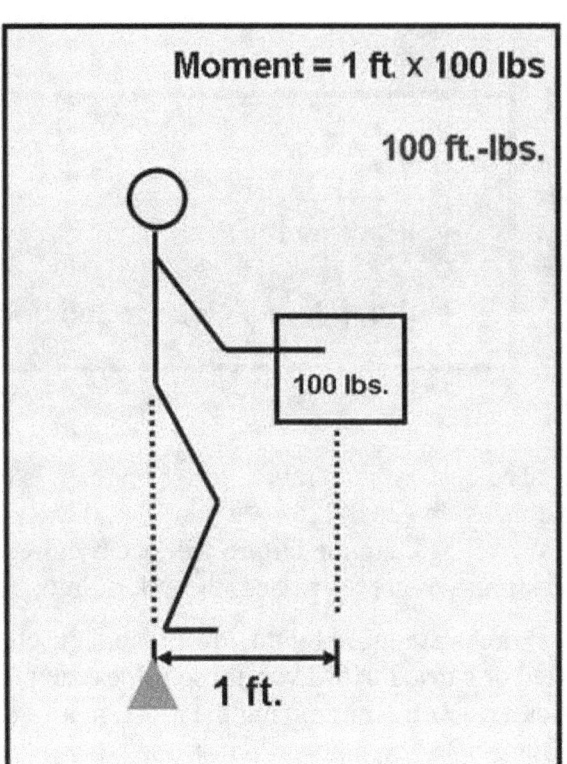

Figure 25. These schematics illustrate how increasing the distance between the worker and the object being lifted increases the overall moment (i.e., torque) for which the back muscles must compensate by expending more force.

Moment Arms

Objectives

To introduce workers to the concept of moments and moment arms

Supplies

A moment-arm simulator (Figure 26)

Three rectangular blocks of equal weight

Step-by-Step Demonstration Method. See Introduction video.

1. Place two of the three blocks on opposite sides of the fulcrum at equal distance from the fulcrum of the moment-arm simulator.

2. The moment-arm simulator should be perfectly balanced.

3. Move one of the blocks to twice the distance from the fulcrum.

4. Note that the "see-saw" will tip towards the block that is furthest from the fulcrum. This occurs because the moment arm (i.e., the distance from fulcrum) is larger for this block. Thus, the moment, or torque, produced by this block is greater than that of the second block.

5. Add a second block to the side of the moment-arm simulator with the shortest distance from the fulcrum.

6. Note that the moment-arm simulator will now balance again, indicating that it is capable of withstanding twice as much force because the moment arm is half as long.

7. Discuss with the audience the point that the weight of the object is not the only consideration in producing forces on the body—as the horizontal distance increases, the resulting moment also increases.

Moment Arms and Lifting

Moment Arms

Take Home Messages

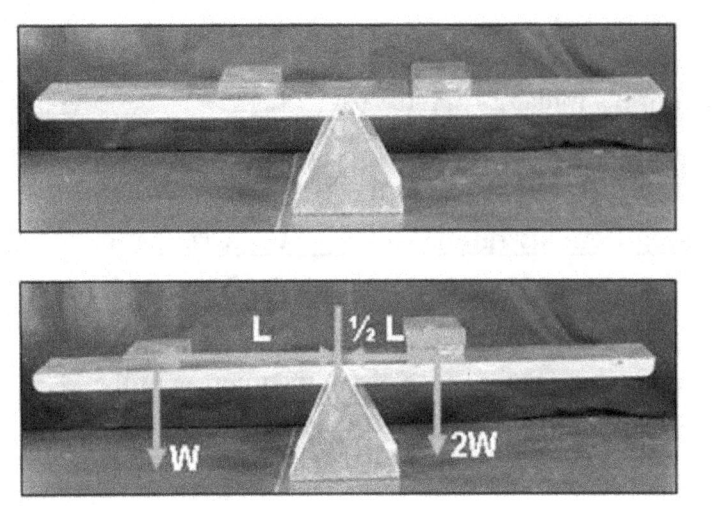

Figure 26. A moment-arm simulator showing that more force/weight (W; arrow indicates direction of force) is needed to balance the "see saw" if the moment arm (L) is shorter on one side of the fulcrum as compared to the other side.

During manual material handling, reduce the moment arm as much as possible by reducing the load on the lower back and keeping the load close to the body.

Design workstations and storage facilities that allow the worker to keep objects close to his or her body when lifting them.

Moment Arms and Lifting

Moment Arms and the Low Back

Objectives

To introduce the effect of moment arm on forces exerted by lower back muscles when lifting or carrying an object

To emphasize that the weight of an object is not the only consideration in determining forces produced by the lower back; the position of the weight of the load in relation to the body also affects the forces and stresses in the low back

To discuss factors that may increase the moment arm and the resulting forces exerted by the low-back muscles

Supplies

A moment-arm simulator made from aluminum (Figure 27)

Spring scale

Metal weights

Step-by-Step Demonstration Method. See Introduction video.

1. Place known weights midway between the fulcrum and the end of the horizontal bench.

2. The moment-arm simulator should be perfectly balanced as the spring scale undergoes loading.

3. Note the force in the spring scale.

4. Move the weights farther away from the fulcrum.

5. Again, the moment-arm simulator should be perfectly balanced; however, the force in the spring scale should increase.

6. Discuss with the audience that the only change made was the moment arm, indicating that weight is not the only factor affecting how much force must be exerted.

7. Relate the spring scale to the low-back muscles, fulcrum to the vertebrae, and weight to an object being carried.

8. Discuss with the audience how the forces exerted by the low-back muscles must increase as an object is moved farther away from the pelvis.

Moment Arms and Lifting

Moment Arms and the Low Back

9. Discuss with the audience the factors that may increase the moment arm when attempting to lift/carry objects—size and shape of the object, existence of a barrier, methods used to complete tasks, or design of workstations.

Take Home Messages

The length of the moment arm and weight of the object both affect the forces exerted by the lower-back muscles.

The size and shape of the object lifted or carried, existence of barriers, and design of workstations are all factors that affect the moment arm of an object being lifted or carried.

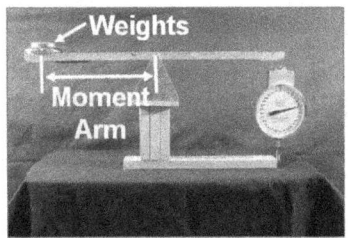

Figure 27. Moment-arm simulator with dial scale showing that, as the moment arm is increased, the resulting force acting on the scale increases.

Moment Arms and Lifting

APPENDIX A: SUGGESTED SUPPLIES

All supplies needed for the demonstrations are included in the following list. Purchasing information is also provided, although most of the supplies can be purchased at hardware stores[1].

- Hand dynamometer (grip type)—evaluates grip strength for multiple grip widths (Figure A-2 A and B).

    o This device measures the hand force generated by a power grip (i.e., where the user curls the fingers towards the palm). The force generated is displayed on a dial or on a digital output. If attempting to locate this device using an Internet search engine (e.g., www.google.com, www.yahoo.com, www.ask.com), the following phrase may be helpful when searching for a vendor—"hand dynamometer grip width." You may also consider adding the keywords "adjustable" or "multiple". This item costs from $225 to $375, depending on the manufacturer and the number of grip widths the dynamometer can evaluate. Among common suppliers of dynamometers, identified from an internet search, are Baseline Tool Company, Medline Industries, and Sammons Preston Rolyan.

    o This device may also be used to evaluate pinch grip, by adjusting the width of the grip to its minimum.

- Hand dynamometer (pinch type)—evaluates pinch grip strength (Figure A-2C).

    o This device measures the force generated by a pinch grip (i.e., where the user presses the thumb against the index finger). The force generated is displayed on a dial or on a digital output. If attempting to find this device using an Internet search engine (e.g., www.google.com, www.yahoo.com, www.ask.com), the following phrase may be helpful when searching for a vendor—"dynamometer pinch grip." You may also consider adding the keywords "strength" or "price." This item costs from $250 to $350 depending on the manufacturer and the maximum force measured by the device. Among common supplier of pinch type dynamometers, identified from an internet search, are Dynatronics, Baseline Tool Company, and Jamar.

- Traditional screwdrivers—varying handle diameters (Figure A-2D).

    o Traditional screwdrivers are designed with a hard, plastic handle. Many manufacturers offer screwdrivers with various handle diameters and working-end size (i.e., a larger handle diameter corresponds to a larger working-end size). However, other manufacturers attempt to keep the handle diameter as large as

---

[1] The National Institute for Occupational Safety and Health does not endorse any manufacturer or supplier of these products. When using the general guidelines provided above, it is the responsibility of the end user to assess the products they intend to purchase and use.

possible. Screwdrivers with different handle diameters, but similar-sized working ends, should be obtained for the demonstrations. These items may be purchased at any standard hardware store for about $20 each.

- Screwdriver—can convert to either an inline or pistol grip (Figure A-2 E and F).

    o Screwdrivers that can be adjusted from an inline to a pistol grip are available at most standard hardware stores. This item costs approximately $50.

- Electric drill—can be held with both hands when additional support is needed (Figure A-2 G and H).

    o Electric drills that have a second handle to allow support from both hands may be purchased at any hardware store. This item costs approximately $100 to $200.

- Needle-nose pliers—with and without a spring-loaded handle (Figure A-2 I and J).

- Needle-nose pliers may be purchased at a hardware store. This item may be purchased for approximately $20.

- Portable electromyography (EMG) device—battery-operated device with audible feedback where the intensity of the sounds produced by the device increases with increased muscle activity (Figure A-2K).

    o When using electrodes placed on the surface of the skin (with tape similar to a Band-Aid), the amount of muscle activity may be measured when a worker is at rest and while performing a task. The amount of muscle activity experienced by the worker is conveyed with an audible sound. As muscle activity increases, so does the intensity of the sound produced by the device. If attempting to locate this device using an Internet search engine (e.g., www.google.com, www.yahoo.com, www.ask.com), the following phrase may be helpful when finding a vendor—"portable EMG audible". This item costs from $300 to $450 depending on the manufacturer. For the purpose of this document, the "pocket ergometer" from AliMed was used; however, NIOSH does not endorse any specific manufacturer[1]. However, a supply of disposable surface electrodes must also be purchased. These typically come in packs of 50, 100, or 1,000. A pack of 100 costs about $10. If using an Internet search engine, the following phrase may be helpful—"EMG disposable surface electrode." Among the common suppliers of electrodes, identified from an internet search, are Nikomed USA, Inc. and Biopac Systems, Inc.

    o The product directions should be followed for the specific device purchased. In general, the device will likely consist of a small box, a cable with three leads at the end, and a package of electrodes. The small box houses the signal processing and output capabilities of the unit. Three electrodes should be placed on the skin above the same muscle or muscle group. Once affixed to the skin, the leads of the cable should be connected to the electrodes. The cable will likely consist of two red leads and one black lead. The black lead should be connected to the electrode that is midway between the other two (Figure A-2K). The red leads should be connected to the electrodes on either side. Once the device is turned on, the audible output may need to be adjusted based on the amount of activity associated with the specific muscle or muscle group being evaluated.

- Standard weights or custom-made blocks (3.5 in x 5 in x 1 in) – four steel blocks weighing approximately 5 lbs each.

    o Standard weights may be purchased at any store that sells supplies for weight training.

    o Custom steel blocks can be constructed from steel stock purchased at most larger hardware stores. Some also offer the service of cutting these items to size for their customers. Other materials may also be used for this task as long as the weight and size of all blocks are uniform.

- Dial scale—must have the ability to attach to objects at either end of the scale (Figure A-2L).

    o This item is best purchased using an online source. Use an Internet search engine (e.g., www.google.com, www.yahoo.com, www.ask.com). The following search phrase may be helpful when finding a vendor—"hanging spring dial scale." This item may cost from $10 to $50, depending on the manufacturer and the maximum force that the device measures. For these demonstrations, a 10-lb capacity or greater is recommended. Among the common suppliers of this device, identified from an internet search, are Detecto, Global Industrial, Calibex, and Salter Brecknell Mechanical Scales.

- Moment-arm simulator that is 40 in long (20 in from fulcrum to end x 5 in width)—must be able to support the weight of the four steel blocks (Figures A-1 and A-2L).

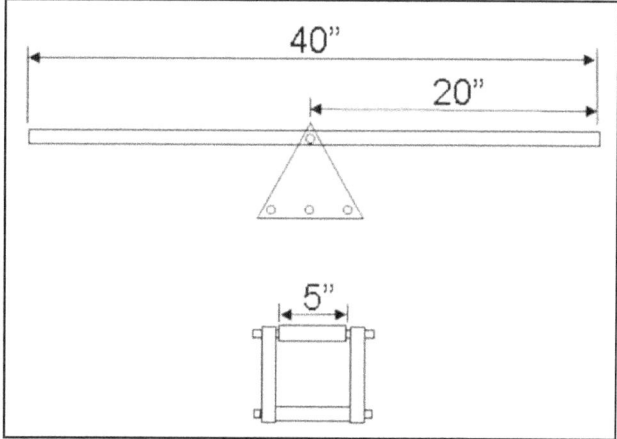

Figure A-1. Dimensions for the moment-arm simulator.

- This item may be custom-made. Suggested dimensions complement the dimensions provided above for the steel blocks. Additionally, this size allows the device to be viewed by all audience members when training is given in a room that is the size of a standard classroom. The "see-saw" should be able to move freely about its fulcrum. Aluminum stock may be purchased at most larger hardware stores; some stores also offer the service of cutting these items to size for their customers. Alternatively, this device may be made from wood at a less expensive cost. Materials other than aluminum may be used. Aluminum was suggested due to its relatively light weight and low cost. You should mark the locations along the see-saw that are 10 inches and 15 inches from the fulcrum as a visual aid for placing the steel blocks during the demonstrations.

- Other similar devices may be purchased for less than $100. An internet search found that common suppliers included Fisher Scientific.

- Pen cap—must be plastic and easy to bend (Figure A-2M)

  - Remove from the top of a pen

- Paper clip—standard size (Figure A-2N)

  - Metal, uncoated paper clips

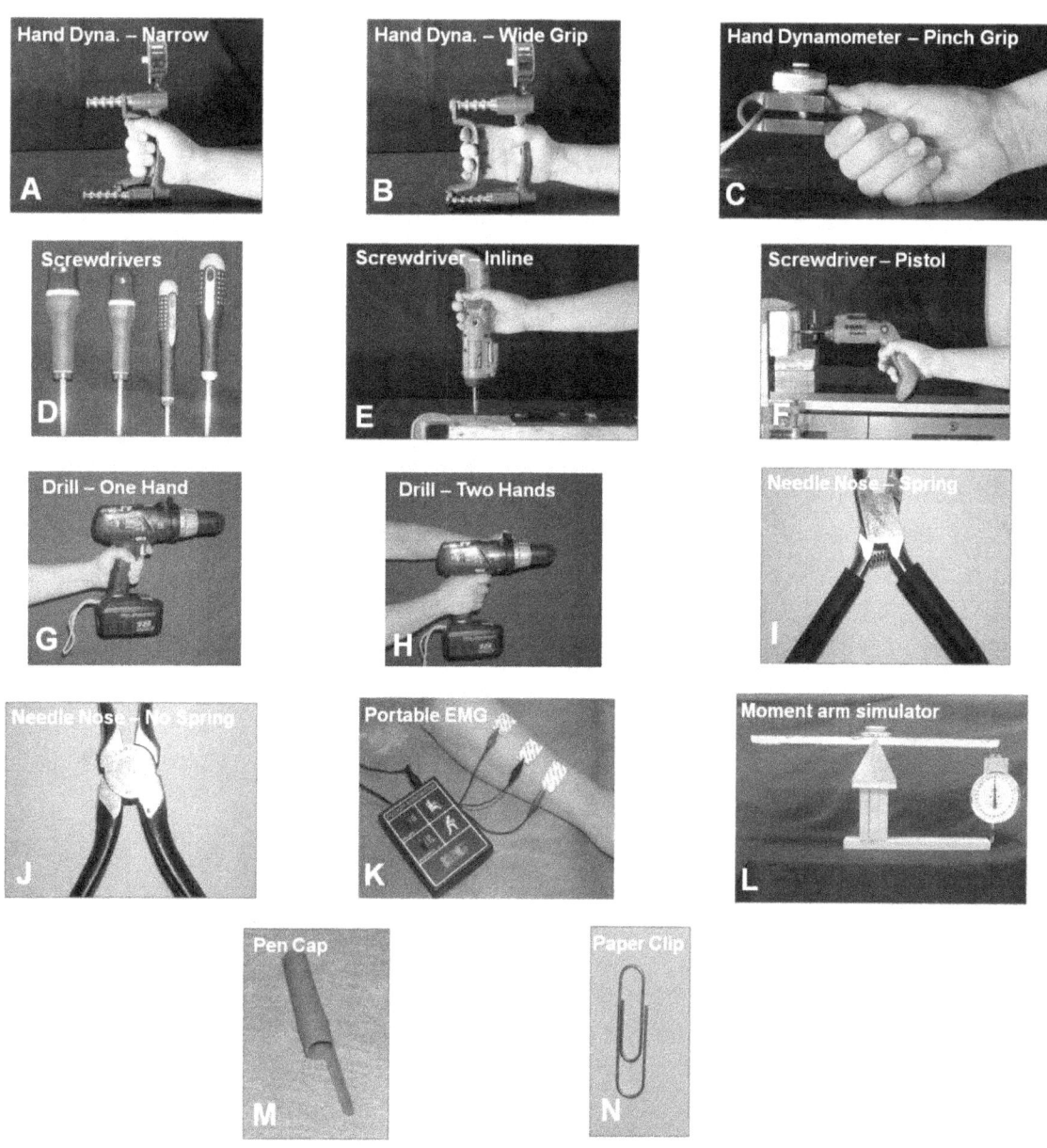

Figure A-2. Examples of the suggested supplies for the demonstrations.

APPENDIX B: USEFUL IMAGES FOR HANDOUTS

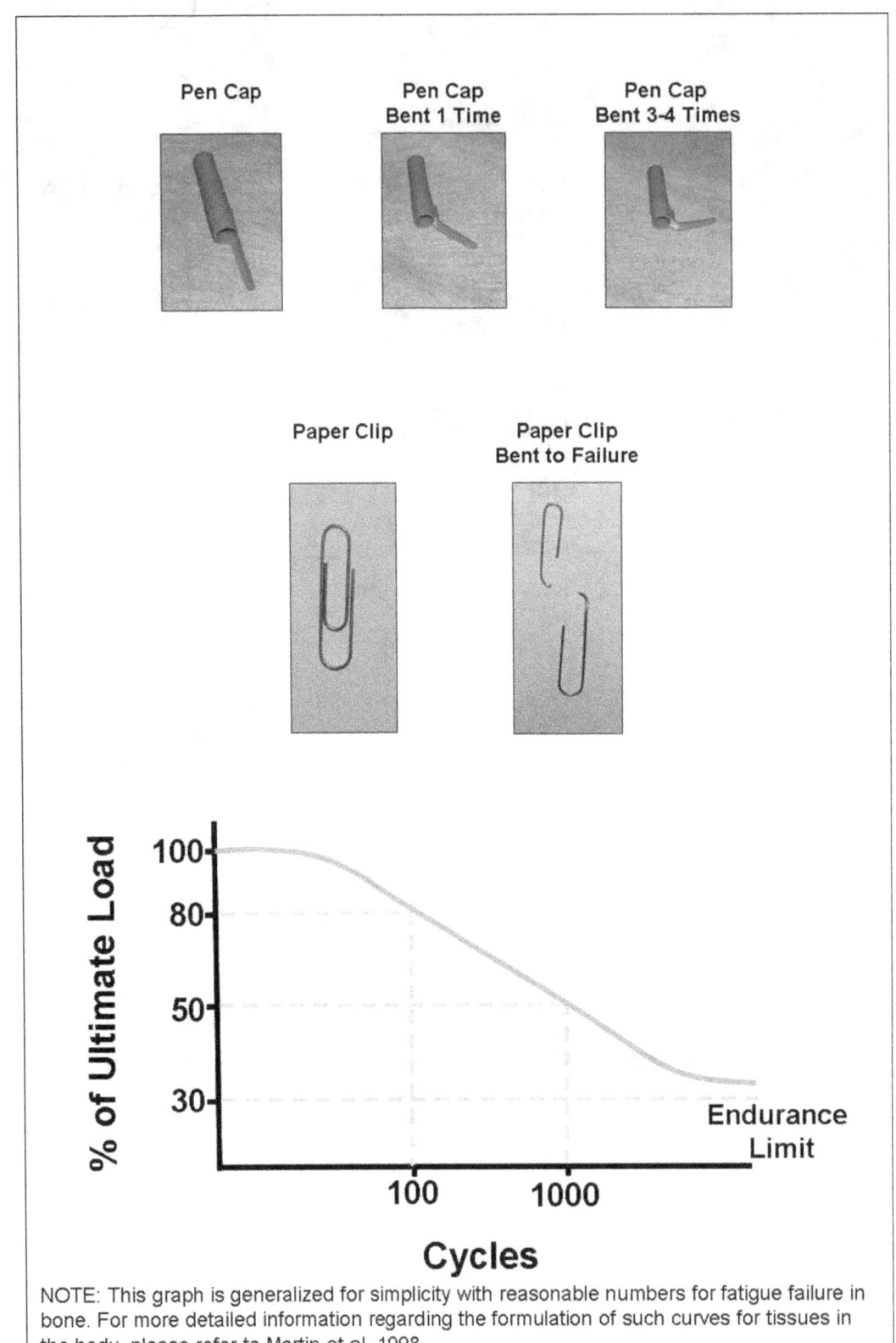

NOTE: This graph is generalized for simplicity with reasonable numbers for fatigue failure in bone. For more detailed information regarding the formulation of such curves for tissues in the body, please refer to Martin et al. 1998.

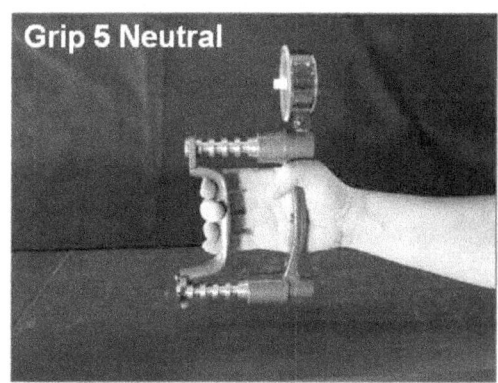
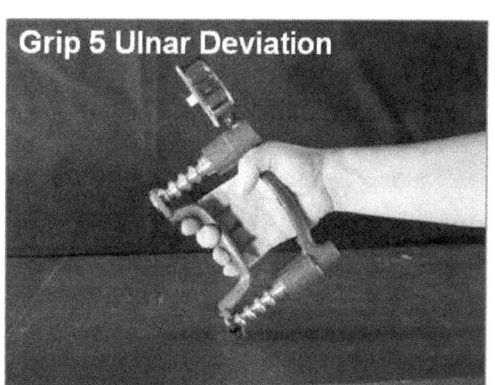
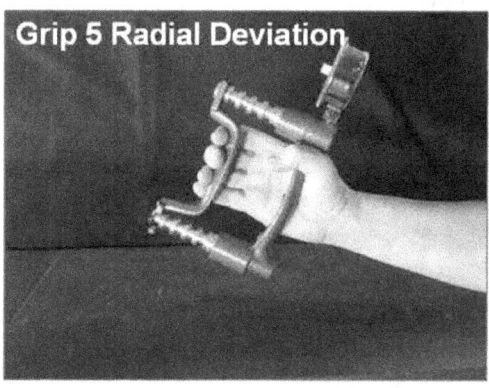
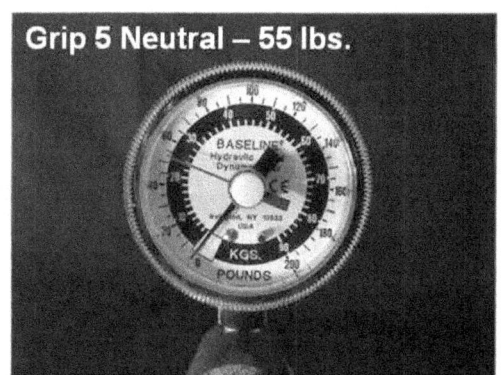
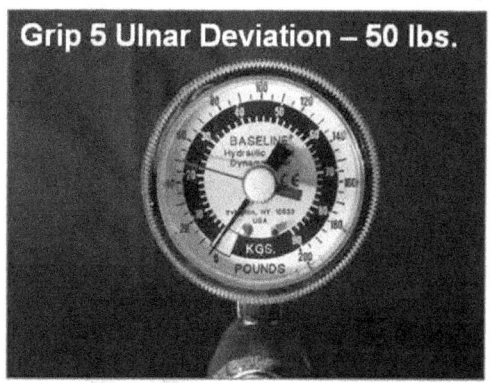
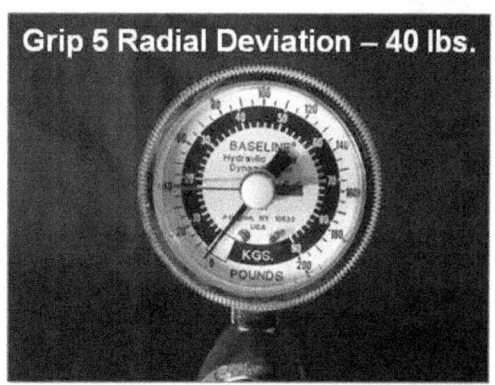

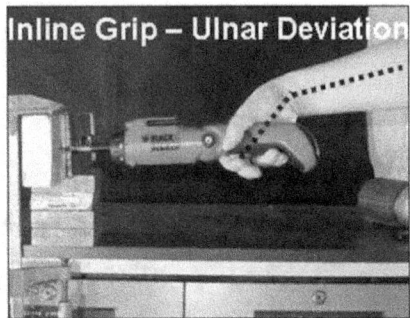
Bent Wrist → Poor Wrist Position

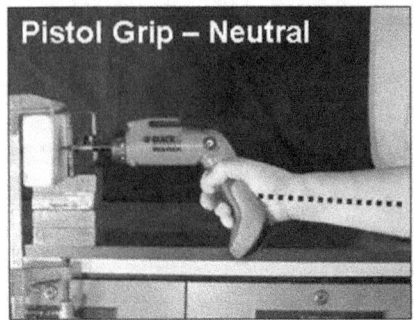
Straight Wrist → Ideal Wrist Position

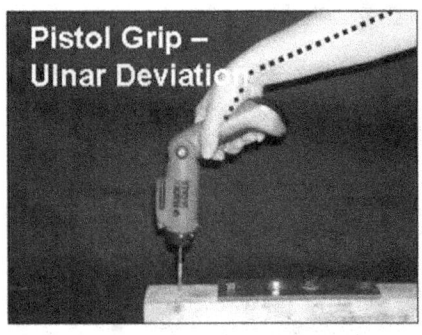
Bent Wrist → Poor Wrist Position

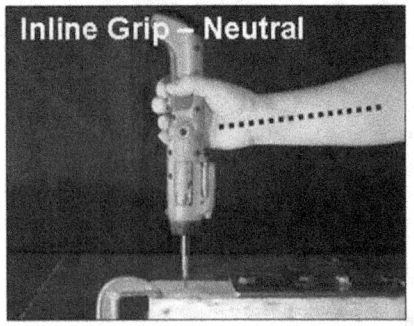
Straight Wrist → Ideal Wrist Position

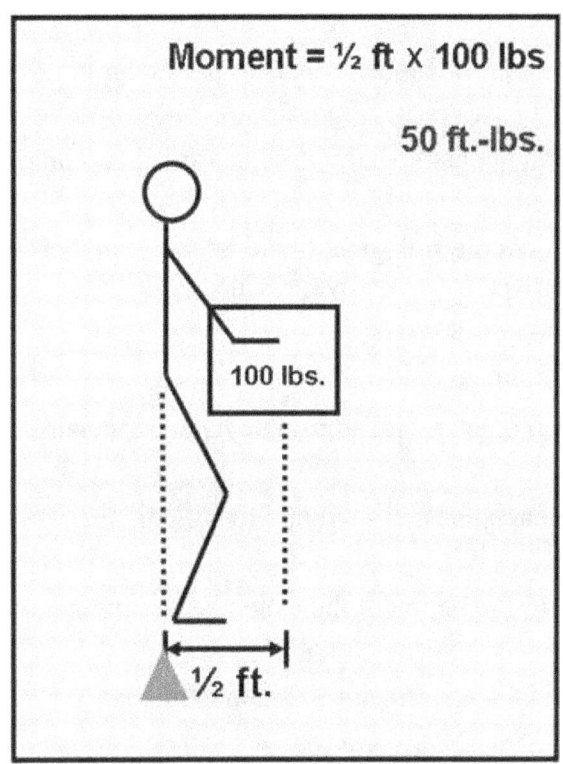
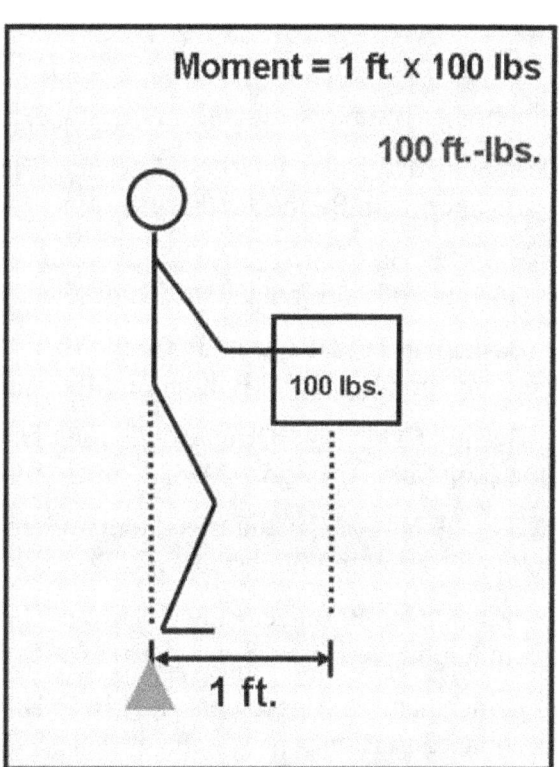

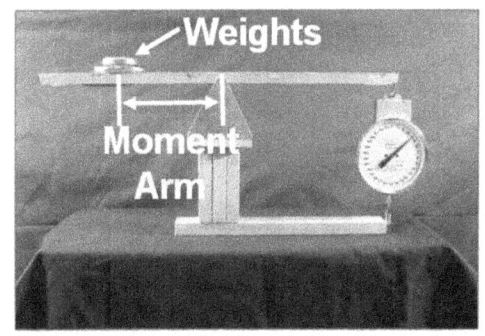
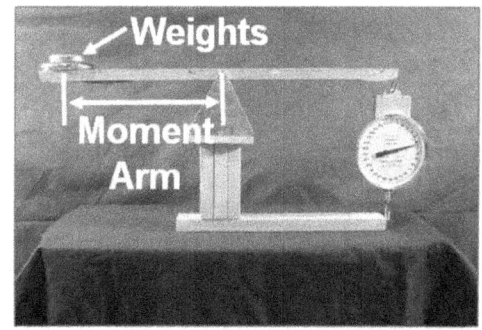

REFERENCES

Adams MA, Bogduk N, Burton K, Dolan P [2006]. The biomechanics of back pain. Edinburgh, Scotland: Churchill Livingstone, p. 316.

Adams MA, Dolan P [1995]. Recent advances in lumbar spinal mechanics and their clinical significance. Clin Biomech *10*:3–19.

Barr AE, Barbe MF [(2004]. Inflammation reduces physiological tissue tolerance in the development of work-related musculoskeletal disorders. J Electromyogr Kines *14*:77–85.

Basmajian JV, De Luca CJ [1985]. Muscles alive: their functions revealed by electromyography, 5th ed. Baltimore, MD: Williams and Wilkins.

Bechtol CO [1954]. Grip test: the use of a dynamometer with adjustable handle spacings. J Bone Joint Surg AM *36*A:820–832.

Bogduk N [1997]. Clinical anatomy of the lumbar spine and sacrum. 3rd ed. Edinburgh, Scotland: Churchill Livingstone, p. 252.

Brinckmann P, Biggemann M, Hilweg D [1989]. Prediction of the compressive strength of human lumbar vertebrae. Spine *14*(6):606–610.

Brinckmann P, Biggemann M, Hilweg D [1988]. Fatigue fracture of human lumbar vertebrae. Clin Biomech *3*(Supplement 1):1–23.

Chaffin DB, Andersson GBJ, Martin BJ [2006]. Occupational biomechanics. 4th ed. New York, NY: John Wiley & Sons, Inc.

Chengalur SN, Rodgers SH, Bernard TE [2004]. Kodak's ergonomic design for people at work 2nd ed. Hoboken, NJ: John Wiley & Sons, Inc., p. 111.

Clarke HH [1966]. Muscular strength and endurance in man. Englewood Cliffs, NJ: Prentice Hall.

Gallagher S, Marras WS, Litsky AS, Burr D [2005]. Torso flexion loads and the fatigue failure of human lumbosacral motion segments. Spine *30*(20):2265–2273.

Jones, R.H. (1974). Unpublished results, Eastman Kodak Company. No other publication information available.

Kumar S [2004]. Muscle strength. Boca Raton, FL: CRC Press, LLC, p. 27.

Marras WS [2008]. The working back: a systems view. New York, NY: John Wiley and Sons, p. 310.

Martin RB, Burr D, Sharkey NA [1998]. Skeletal tissue mechanics. New York, NY: Springer-Verlag.

NIOSH [2008]. Ergonomics and risk factor awareness training for miners. By Torma-Krajewski J, Steiner LJ, Unger RL, Wiehagen WJ. Cincinnati, OH: U.S. Department of Health and Human Services, Public Health Service, Centers for Disease Control and Prevention, National Institute for Occupational Safety and Health, DHHS (NIOSH) Publication No. 2008-111, Information Circular 9497.

NIOSH [2004]. Easy ergonomics: a guide to selecting non-powered hand tools. By Hight R, Schultz K, Hurley-Wagner F, Feletto M, Lowe BD, Kong YK, Waters T. Cincinnati, OH: U.S.

Department of Health and Human Services, Centers for Disease Control and Prevention, National Institute for Occupational Safety and Health, DHHS (NIOSH) Publication No. 2004–164.

NIOSH [1994]. Applications manual for the revised NIOSH lifting equation. By Waters TR, Putz-Anderson V, Garg A. Cincinnati, OH: U.S. Department of Health and Human Services, Centers for Disease Control and Prevention, National Institute for Occupational Safety and Health, DHHS (NIOSH) Publication No. 94–110.

Ozkaya N, Nordin M (1999). Fundamentals of biomechanics: equilibrium, motion, and deformation. 2nd ed. New York, NY: Van Nostrand Reinhold Company.

Sanders MS, McCormick EF [1993]. Human factors in engineering and design. New York, NY: McGraw-Hill, Inc.

Silverstein BA, Stetson DS, Keyserling WM, Fine LJ [1997]. Work-related musculoskeletal disorders: comparison of data sources for surveillance. Am J Ind Med *31*(5):600–608.

Silverstein MA, Silverstein BA, Franklin GM [1996]. Evidence for work-related musculoskeletal disorders: a scientific counterargument. J Occup Environ Med *38*(5):477–484.

Waters TR, Putz-Anderson V, Garg A, Fine LJ [1993]. Revised NIOSH equation for the design and evaluation of manual lifting tasks. Ergonomics *36*(7):749–776.

www.ingramcontent.com/pod-product-compliance
Lightning Source LLC
Chambersburg PA
CBHW081855170526
45167CB00007B/3023